PRACTICAL GUIDE TO DIAGNOSING
STRUCTURAL MOVEMENT IN BUILDINGS

PRACTICAL GUIDE TO DIAGNOSING STRUCTURAL MOVEMENT IN BUILDINGS

Second Edition

Malcolm Holland

WILEY Blackwell

This edition first published 2023
© 2023 John Wiley & Sons Ltd

Edition History
John Wiley & Sons Ltd (1e, 2012)

Registered Office(s)
John Wiley & Sons, Inc., 111 River Street, Hoboken, NJ 07030, USA
John Wiley & Sons Ltd, The Atrium, Southern Gate, Chichester, West Sussex, PO19 8SQ, UK

Editorial Office
9600 Garsington Road, Oxford, OX4 2DQ, UK

For details of our global editorial offices, customer services, and more information about Wiley products visit us at www.wiley.com.

Wiley also publishes its books in a variety of electronic formats and by print-on-demand. Some content that appears in standard print versions of this book may not be available in other formats.

A catalogue record for this book is available from the Library of Congress
Paperback ISBN: 9781119898726; epub ISBN: 9781119898740; ePDF ISBN: 9781119898733

Cover image: © bridixon/E+/Getty Images
Cover design by Wiley

Set in 10/12pt Minion Pro by Integra Software Services Pvt. Ltd., Pondicherry, India
Printed and bound by CPI Group (UK) Ltd, Croydon, CR0 4YY

C9781119898726_190224

Contents

Introduction

It is often the layman's first reaction when cracking is observed in a building, that it must be the foundations and that it is serious. This is not true. In the vast majority of cases, it is not subsidence or settlement of the foundations – and in most cases cracks do not indicate a serious defect. So when analysing cracks, it is essential to always keep an open mind. A good rule of thumb for the beginner is to try to find what has caused the crack other than foundation movement. Only when all other possibilities have been ruled out, consider whether it is foundation movement. In my experience (teaching university students and graduate surveyors), it is very difficult to instil this discipline. There is a great temptation to jump to a conclusion or to shortcut the process of analysis.

Many people, including surveyors, are nervous about diagnosing cracks. This is understandable, as the consequences of getting it wrong can be potentially onerous. It is a subject that is difficult to teach in the lecture theatre and there may be little or no time for field experience within an academic syllabus at university. Most text books on the subject are aimed more towards Engineers or Building surveyors, who specialize in this area. Necessarily such books are very technical in nature. This book is not aimed at the experienced engineer or surveyor. It is primarily intended for the relatively inexperienced surveyor, engineer, undergraduate or even a competent layman. It is intended as a practical guide or as on-site manual. The intention is to remain as un-technical as possible. It avoids references to regulations, digests and other technical sources. These can be found elsewhere in other books on the subject. The book is concerned with identification and diagnosis not the detailed specification of remedial work.

The intention is to know, with a reasonable level of confidence, when movement is potentially serious or not. To know when it is necessary to call in the more experienced or qualified professional to deal with it.

The purpose of this book is to show that by understanding one simple first principle and by following a simple methodology, the vast majority of cracks, probably as many as nine out of ten cracks, can be diagnosed in just a few minutes.

By looking at the cracks in a little more detail and by understanding the factors that distort crack patterns, this diagnosis rate can be raised even further.

Couple this with a reasonably good knowledge of building construction and the key features of the most common causes of cracking; and almost all cracks can be diagnosed relatively quickly and with a high degree of confidence. There will however always be some cracks that cannot be diagnosed from a single inspection. Inevitably, when movement first starts to develop, the evidence can be insufficient to reach a conclusion. In some cases movement will have to be monitored for a period of time. In other cases the only way to obtain enough evidence to make a diagnosis, might involve opening up the structure of a building or carrying out excavations to expose the foundations. Material or sub soil samples may need to be taken for testing. A building may need to be monitored for a period of time to make a diagnosis or to confirm a preliminary opinion. The number of cases where such action is necessary is likely to be small providing that the basic principles are understood and applied in an objective manner.

This may seem too easy and too good to be true but why should it not be true? Cracks are caused by a simple physical process and the physics always acts in the same way. It is a 'law' of physics; there is no hidden agenda, no politics; just a simple physical process. Tension in a material or structure will act to try to pull it apart.

When the tension is sufficiently large in relation to the strength of the material it will pull it apart and cause a crack. It is a simple physical process in the same way that adding heat to a liquid causes a physical process of temperature rise and when enough heat is added in relation to the boiling point of the liquid it will cause it to boil. It will always give the same result.

This book provides a methodology by which cracks and movement in buildings can be diagnosed.

The book is in four parts. The first part describes the first principles. The second part of the book contains 'swatches' which describe the key features of the most common forms of movement in buildings and the crack patterns associated with them. This part covers movement and cracks NOT caused by ground or foundation movement. The third part of the book contains similar 'swatches' giving the key features relating to movement caused by ground or foundation problems. Part four describes the techniques used to repair damage cause by movement and to arrest further movement.

By applying the FIRST PRINCIPLES and then referring to the 'swatches', there should be a high probability that the correct diagnosis can be reached.

The methodology contained within this book will not only help to derive the correct diagnosis but it will also demonstrate a process to show and record how the diagnosis was reached. When giving advice to a third party, the ability to demonstrate a proper methodology, a chain of thought and logical process is critically important should it ever become a negligence claim. When advising a third party it is also imperative to make clear that any diagnosis is an opinion, not a guarantee. In theory an opinion can turn out to be wrong without negligence automatically occurring provided that a correct methodology has been followed and that other reasonably competent persons would have arrived at the same opinion. In the light of previous case law some might argue this to be an over-optimistic view for a professional working in this field to take; but objectively that should be the case.

So it is absolutely critical that one understands the FIRST PRINCIPLES before moving on to the second part of this book. It is straightforward and relatively simple but if on first reading you do not completely grasp it, read it again. Do not be tempted to short cut the process by simply looking at the examples or photographs in the book and finding one that looks similar.

And finally, good luck.

List of Figures

Acknowledgements

I would like to acknowledge my mother-in-law, Mrs Jill Porter, and my ex-boss, Martin Brown FRICS, for their help with editing and proofreading.

I would also like to acknowledge all those people who have freely passed on their knowledge and experience to me throughout my career. I hope that I can pass on their baton.

Part 1

First Principles

1.1

First Principles

> *Most traditional building materials are relatively weak in tension, when compared to their compressive strength. If a building is distorted, by whatever force, some parts of it will be stretched. Cracking is likely to occur at right angles to the force that caused the stretching. By imagining arrows at right angles to a crack, it is possible to determine the direction of movement. The direction of the movement is usually directly related to its cause. There are, however, always some cracks that cannot be diagnosed quickly by a simple visual inspection.*

The overriding first principle that one must understand when diagnosing cracking is that the materials we are dealing with, bricks and concrete, are weak in tension.

STEP ONE

Brickwork and most materials crack when pulled apart in tension. Tension is caused by elongation (Figure 1.1.1).

Practical Guide to Diagnosing Structural Movement in Buildings, Second Edition.
Malcolm Holland.
© 2023 John Wiley & Sons Ltd. Published 2023 by John Wiley & Sons Ltd.

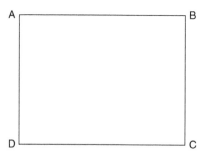

Figure 1.1.1   Diagonals of equal length.

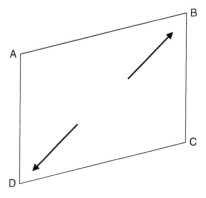

Figure 1.1.2   Diagonal B–D stretched.

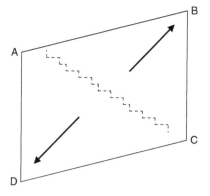

Figure 1.1.3   Crack at right angles to tension.

Imagine a square or rectangle of material – A, B, C, D. Think of this as the front elevation of a building or a panel of part of a building.

Within normal tolerances buildings are built square and plumb. In a square or rectangle, the diagonals are the same length.

Please measure them. A–C is the same length as B–D (Figure 1.1.2). Please write the measurements down.

If the left-hand side settles, the diagonal A–C is shortened. It is put into compression (Figure 1.1.3). The diagonal B–D is lengthened. It is put into tension.

Please measure B–D and write the measurements down. You will see that it is longer than it was. It is this stretching that is important as it creates tension.

The tension force pulls the panel apart, in tension. If it was brickwork, it would pull it thus. The crack would be perpendicular to the elongation. THE CRACK IS AT A RIGHT ANGLE TO THE TENSION.

This is always the first point to look at when observing cracks. The movement is at right angles to the crack. The building has either moved up to the right-hand side or down to the left-hand side. There are very few reasons for upward movement and this possibility can quickly be assessed and usually dismissed. Buildings are heavy and gravity pulls them down. It is the downward arrow that is usually significant.

Example 1

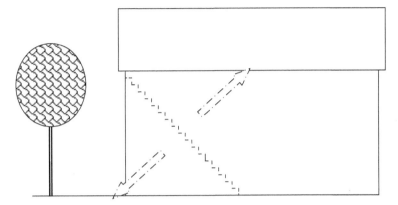

Figure 1.1.4 Imaginary arrows at right angles to tension.

If one imagines arrows at right angles to the crack, as shown dashed in the diagram in Figure 1.1.4, then one can see that the movement is either down to the left-hand corner or up to the right-hand side. The arrows point to where the movement is and this is usually pointing at the defect. In this case one would look down to the left and there is the tree. Alternatively, the movement could be up to the right-hand side. This is very unlikely and in this case, there is nothing that could possibly cause movement up to the right. The solution is therefore reasonably obvious. In the absence of other evidence, the movement is likely to be caused by the tree.

There are some instances where upward movement can occur in buildings; the most common being clay heave or corrosion of steel fixings and wall ties. Although upward movement is far less likely than downward movement the possibility of upward movement should be assessed before dismissing the possibility.

Example 2

It is quite common to see diagonal cracking above an opening in a brick wall, similar to that shown in the diagram in Figure 1.1.5. The cracking forms a triangle running diagonally through the brick courses at around 45° from the support points at either side.

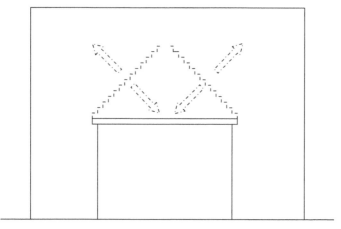

Figure 1.1.5 Imaginary arrows of tension intersecting on a supporting beam.

Once again, imagine arrows at right angles to the cracks as shown in the diagram. In this case two of the imaginary arrows at right angles to the cracks intersect on the lintel – good evidence. In fact, if two arrows intersect at the same point, it is almost certainly the position of the defect. The two arrows pointing up contradict each other, one up to the left and one up to the right. The upward contradicting arrows can be recognized as redundant and ignored.

The solution is again obvious. The arrows converge on the lintel (beam) and the movement is deflection of the lintel (beam) supporting the masonry above the opening.

Example 3

Look at a very similar situation in Figure 1.1.6.

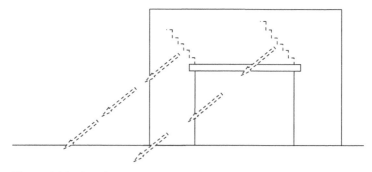

Figure 1.1.6 Imaginary arrows of tension point to left-hand pier.

Imagine arrows at right angles to the cracks. There is no reason for upwards movement here so the arrows pointing up to the right are dismissed. The other arrows show movement down to the left. The arrows running from the right-hand side first intersect on the lintel. Could it be lintel deflection? If it was lintel deflection the arrows from the left-hand crack would also point to the lintel, as they did in the previous example. These arrows go down to the left-hand side. If it is not the lintel, continue the arrows from the right-hand side down, and they point to the next area, being the left-hand pier to the side of the opening. There are two cracks, and both sets of arrows point down to the left-hand side. The cause of the movement is there, in the left-hand pier. Having established where the movement is, we now have to work out why.

In this case study there is not enough information to work out why. The settlement in the left-hand pier could simply be caused by load concentration. It could be poor foundations or poor soil-bearing capacity. In real life, there might be a tree or a defective drain nearby.

We would have to collect that information sequentially on site, in order to work out which would be the most probable. If there was no tree or no drain, and the movement was significant in magnitude, we may have to arrange for some trial pits to be dug, in order to determine the nature of the foundations and the subsoil. Only then might it be possible to make the final diagnosis. There will inevitably be cracks that cannot be definitively diagnosed on the basis of one visual inspection.

1.2

Crack Patterns and Cracks

> *By following the simple principles of tension and compression, most cracks can be diagnosed quickly. There are, however, a number of things that can distort crack patterns. By understanding the factors that distort the shape and direction of a crack, a more reliable diagnosis can be made.*

By the simple application of the First Principles in STEP 1 most cracks can be diagnosed within a few minutes. Probably around nine out of ten cracks can be diagnosed almost immediately. By sketching a diagram similar to those in the examples, denoting the building, pattern of the cracks and arrows of tension at right angles to the crack; the position of the movement will usually be obvious.

There are, however, several factors that can distort the way cracks appear. To improve the success rate of diagnosis these factors need to be understood.

Practical Guide to Diagnosing Structural Movement in Buildings, Second Edition.
Malcolm Holland.
© 2023 John Wiley & Sons Ltd. Published 2023 by John Wiley & Sons Ltd.

1.3

Rotational Movement

When buildings subside, they are unlikely to descend evenly. As one part goes down in relation to another part, there is a 'hinge' effect. The hinge effect causes rotation to occur. This increases the horizontal displacement of the crack with height. Cracks created by subsidence or settlement are likely to taper, being narrow at the base and wider at the top.

When a building is affected by subsidence, it rarely drops straight down. In order for a building to drop straight down, the subsidence would have to be evenly distributed around the entire area of the building. That is not usually the case. A defect in one area is the usual cause of subsidence, which then spreads from the epicentre of the defect.

For example, Figure 1.3.1 shows a typical subsidence crack, caused by clay shrinkage and the proximity of a tree. The tree extracts moisture from the clay soil. The clay shrinks in this location. The foundations nearest to the tree begin to subside.

The brickwork is bonded together. As a result, there is resistance to the movement. This drag causes a 'hinge' effect, and the wall rotates. As it rotates it moves out of vertical at the top. The effect of this is to increase the amount of horizontal displacement at the top of the wall. This creates one of the key features of subsidence cracking. That is the cracks are wider at the top than they are at the bottom.

Practical Guide to Diagnosing Structural Movement in Buildings, Second Edition.
Malcolm Holland.
© 2023 John Wiley & Sons Ltd. Published 2023 by John Wiley & Sons Ltd.

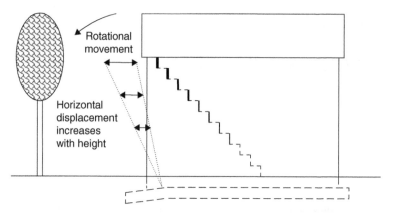

Figure 1.3.1   Rotational movement.

The degree to which they get wider will vary in each case, depending on how much rotational movement there is, in relation to vertical movement. If, for example, the walls have been built outside the centre line of the foundations, creating eccentricity, the rotational horizontal displacement is likely to be greater than it would be, if the wall had been built correctly on the centre line of the foundations.

Contra Rotational Movement

> *Rotational movement of a building often lifts off the brickwork at the top. Imagine a 'see saw' effect. This lifting creates cracking, that is contrary to the direction of the main movement. This cracking has to be identified so that it can be excluded from the diagnosis. This effect is demonstrated in a graph paper exercise.*

Rotational movement sometimes has another effect on the crack pattern. It sometimes creates cracks, which are contrary to the overall pattern of the movement. Such contrary movement must be recognized and dismissed.

Figure 1.4.1 shows a typical settlement pattern crack roughly at about 45°. There is some variance of the angle, through window openings. By sketching in imaginary 'arrows of tension', it can be seen that the main movement is down to the left. At the very top of the wall however, the crack runs in the opposite direction. An imaginary arrow of tension through the crack would point in the opposite direction of the main movement.

Practical Guide to Diagnosing Structural Movement in Buildings, Second Edition.
Malcolm Holland.
© 2023 John Wiley & Sons Ltd. Published 2023 by John Wiley & Sons Ltd.

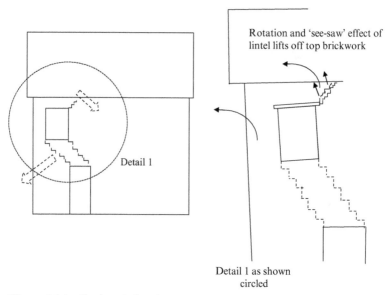

Rotation and 'see-saw' effect of lintel lifts off top brickwork

Detail 1

Detail 1 as shown circled

Figure 1.4.1   Contra rotational movement.

This can be explained by the rotational effect. Rotational settlement at the bottom left-hand corner moves the wall out of vertical. As it tips down at one end, it lifts at the other. This is a bit like a see-saw. The rigid lintel over the window acts as a lever. It lifts the brickwork off at upper level, giving a crack pattern in the opposite direction to the main crack. The rotation creates a crack that is contrary to the main movement at the top of the wall. There is no official title to describe this phenomenon. For the purpose of this text it is referred to as 'contra rotational movement'.

In the example above, it was the lintel that had sufficient tensile strength to be able to act as a lever, to lift the brickwork. A similar effect often occurs where the internal leaf of a cavity wall, or an internal frame, is able to transmit tensile load. For example, cavity walls built with an inner leaf of reinforced cast in situ concrete. Other examples would be timber-framed or steel-framed buildings.

In order to clarify how contra rotational movement occurs try the following exercise.

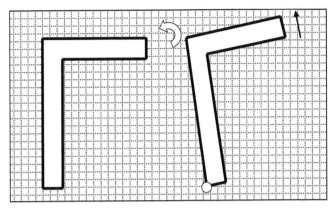

Figure 1.4.2 Graph paper exercise of counter rotational movement.

Cut out an 'L' shaped piece of card and place it upside down on a piece of graph paper as shown in Figure 1.4.2. Pin the bottom left-hand corner. Rotate the card around anticlockwise. As the shape rotates to the left, the top right-hand corner lifts up, above its original position.

Weak Routes

> *Openings in walls for windows and doors reduce the strength of brick-work. The direction of a crack may be distorted by the relative position of openings within a wall or elevation of a building. The tension force will cause failure along the weakest route. Depending on the relative position of the openings, the angle of the crack might be made more horizontal or more vertical, than would otherwise be the case.*

Buildings usually have openings in the walls, for windows and doors. Where there is an opening in a wall, the strength of the brickwork is reduced. The tension force will overcome the ability of the brickwork to resist it, along the line of the weakest route. Depending on the position of the openings, this will alter the angle of the crack pattern.

Although the angle of the cracks shown in Figures 1.5.1, 1.5.2 and 1.5.3 are different, they are, after allowance for weak routes, the same crack. In Figure 1.5.1, the wall is just plain brickwork, with no openings. The strength of the wall is reasonably uniform. A subsidence crack will usually be at about 45°, and will follow the coursing through the bricks and mortar joints.

In Figure 1.5.2, there are openings in the wall. The openings are weak points. The crack will therefore run up the sides of the openings, being a door at ground-floor level and a window at first-floor level, in this case.

Practical Guide to Diagnosing Structural Movement in Buildings, Second Edition.
Malcolm Holland.
© 2023 John Wiley & Sons Ltd. Published 2023 by John Wiley & Sons Ltd.

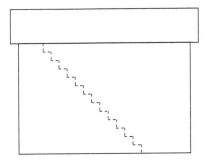

Figure 1.5.1 Typical subsidence crack at about 45°.

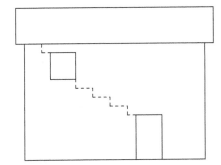

Figure 1.5.2 Angle reduced below 45° due to position of openings.

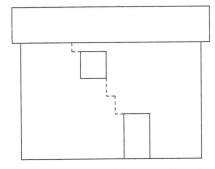

Figure 1.5.3 Angle steeper above 45° due to position of openings.

Between the openings, the crack will exploit the weakest route through the area where there is the least amount of brickwork, to resist the tension. This will be a diagonal line between the openings. If the openings are positioned so that line is below 45°, the resulting crack will be below 45°.

In Figure 1.5.3, the same principle applies. In this example the openings are positioned in such a way that the most direct route of least resistance is steeper than 45°.

When we sketch in imaginary 'arrows of tension' at right angles to the cracks on the three diagrams, we would get different angles. The shallow crack, shown in Figure 1.5.2, would give a more vertical angle of tension. The steeper crack, shown in Figure 1.5.3, would give a more horizontal angle of tension. Provided that we can recognize that the crack patterns have been distorted by the position of the openings, we can adjust the angle of the 'arrows of tension', back to where they should be. This is where professional judgement and experience come into play. The adjusted arrows of tension will no longer be at right angles to the actual crack. They will have been adjusted as required, (shallower or steeper), to compensate for the crack pattern being distorted by the position of the openings.

Weak routes can also make expansion crack patterns look quite similar to subsidence patterns. In order to be able to differentiate between the two it is necessary to look not only at the overall crack pattern, but the crack itself.

Figure 1.5.4 shows a 'textbook' case of cracking associated with subsidence, typically stepped at about 45°. Figure 1.5.5 shows a textbook example of an expansion crack pattern, typically vertical.

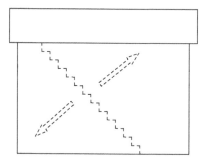

Figure 1.5.4 Typical subsidence crack.

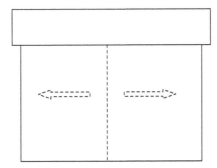

Figure 1.5.5   Typical expansion crack.

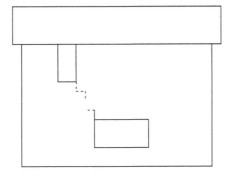

Figure 1.5.6   Expansion crack distorted by openings.

More detailed examples of these crack patterns will be given in Part 2 and Part 3 of this book, together with key features to aid diagnosis.

If there are openings in the walls, creating a weak route, this can distort the angle of the crack, as previously explained. Figure 1.5.6 shows an expansion crack that has been distorted by the position of openings, which have created a weak route.

The overall appearance of the crack is angled. This could easily be confused with subsidence. In order to differentiate between the two, it is necessary to look at the crack itself and note the degree of displacement, both vertically and horizontally.

A subsidence crack is likely to have both horizontal and vertical displacement. The width of the crack measured horizontally is likely to be a bit wider than the vertical width of the crack. This is because of rotational movement makes the cracking wider horizontally. Also, gravity tends to close up the vertical displacement, as the bricks above the crack slip down on the bricks below the crack. The crack will also generally be wider at the top than the bottom, as a result of rotational movement. An expansion crack will be nearly uniform in width and all the displacement will take place horizontally. The only movement on the bed joint will be where the bricks have been slid across from side to side. The crack on the bed joint will be hairline and the bricks and mortar will, in effect, still be touching.

Figure 1.5.7 shows part of a typical crack that has been caused by subsidence. The crack pattern is near to 45°. The rotational effect increases the amount of horizontal displacement, in relation to the amount of vertical displacement. The amount of rotation will vary from case to case, and this is an area where 'professional judgement' must be made in some cases. On the less obvious cracks, there is, unfortunately, no replacement for experience.

In Figure 1.5.7, there is a small thumbnail sketch of one part of the crack, shown ringed. This example suggests that a 3 mm vertical displacement might result in a 5 mm horizontal displacement. This is purely for example purposes. The amount could vary depending on individual circumstances, but the figures suggested are considered to be quite representative of what would be expected, in a real case.

Compare this to the typical expansion crack shown in Figure 1.5.8. The expansion crack pattern in this example has been distorted into a stepped crack, by the position of window openings in the wall. This could make it look similar overall to the subsidence crack shown in Figure 1.5.7. A close-up detailed inspection of the crack, however, will reveal that all the displacement and separation is horizontal. On the bed joint there is no separation, and the crack is merely hairline. The hairline cracking on the bed joint is a shearing action caused by horizontal sliding of the bricks over those beneath.

There is a thumbnail sketch of a small area of the crack shown circled. The figures are again for example purposes only. The amount of displacement would vary depending on the co-efficient of expansion of the masonry, and the temperature variation to which the wall was exposed. The figures are, however, considered to be quite realistic. The coefficient of expansion is a measure of how much any material will expand with each degree change in temperature. Each material will have a different rate of expansion. Steel, for example, has a high coefficient of expansion, when compared to brickwork. Concrete also has a higher coefficient of expansion than brickwork.

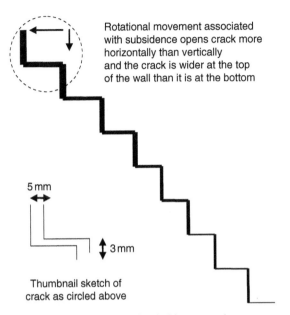

Rotational movement associated with subsidence opens crack more horizontally than vertically and the crack is wider at the top of the wall than it is at the bottom

5 mm

3 mm

Thumbnail sketch of crack as circled above

Figure 1.5.7   Appearance of a typical subsidence crack.

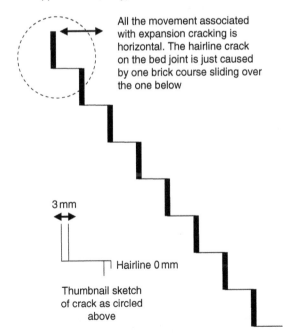

All the movement associated with expansion cracking is horizontal. The hairline crack on the bed joint is just caused by one brick course sliding over the one below

3 mm

Hairline 0 mm

Thumbnail sketch of crack as circled above

Figure 1.5.8   Appearance of a typical expansion crack that has been distorted out of vertical by the position of window openings.

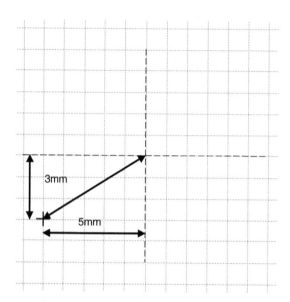

Displacement measured from subsidence crack plotted
on graph paper.

Figure 1.5.9   Subsidence plotted on graph paper.

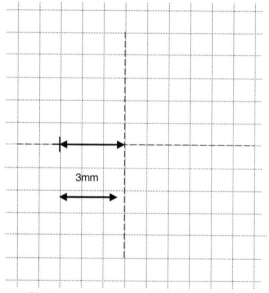

Displacement from expansion crack plotted on
graph paper.

Figure 1.5.10   Expansion plotted on graph paper.

In order to demonstrate this in another way, it would be possible to measure the horizontal and vertical displacement on site, and then plot the angle of movement (angle of imaginary tension lines) on a piece of graph paper.

For the mathematically minded the angle of movement could also be calculated from the measurements by using trigonometry.

If a vertical and horizontal axis is drawn on a piece of graph paper, the displacement can be plotted to scale. Figures 1.5.9 and 1.5.10 are at a scale of one graph square to 1 mm. In practice 1 cm to 1 mm displacement, or 2 cm to 1 mm displacement, would be practical.

The displacement, 3 mm down and 5 mm to the left of the starting centre axis point, is plotted (see Figure 1.5.9). A line drawn from the centre axis to the plotted point indicates the direction of movement, which is the same as the 'imaginary arrows of tension'. The plotted line is not exactly at 45°. As previously mentioned, this can be adjusted as a matter of professional judgement to compensate for the rotational effect. The movement can be seen to be down to the left.

In Figure 1.5.9, the 3 mm displacement to the left is plotted. The hairline vertical displacement is virtually zero. The plotted point is virtually on the horizontal axis line. By connecting the centre axis point to the plotted point, we can see that the direction of movement is all horizontal. Horizontal movement is a typical feature of expansion cracking.

In this example, the overall appearance of the expansion crack has been made sloped, by the position of window openings. When the actual displacement is plotted however, it reveals only horizontal movement. We are therefore able to adjust the 'imaginary lines of tension', from the stepped pattern of the crack overall, to the real direction of tension, which is horizontal. By looking at the appearance of the crack itself, as well as the overall pattern, it is possible to distinguish between a subsidence crack, running at an angle, and a sloping expansion crack, that has been distorted to run at an angle.

1.6

Load Distribution

> *Forces created in a building, or element of a building, are able to move more lightly loaded parts to a greater degree than the more heavily loaded parts. The load on a building is not evenly distributed. Loads are greater at the bottom of the wall than the top. Load concentrations occur at the support points over openings. Loads are often concentrated through narrow piers of brickwork. A simple line diagram can be used to imagine the 'flow' of the load through the building and around openings.*

The other major factor influencing the shape of a crack is the load distribution. Not all parts of a building are loaded equally. There are load concentrations at the sides of openings where lintels and beams transfer the weight from above. Immediately below window openings there is no load on the brickwork, other than its own self-weight.

At the top of a wall there is little weight, only that of the roof. The weight on the brickwork increases progressively towards the bottom, as it has to support the entire load from above. Given an equal force, it will be easier for that force to move areas that are lightly loaded, rather than those that are heavily loaded. This can again cause distortion of a crack pattern to some degree. For example, in the previous section, expansion cracks were compared with subsidence cracks. Expansion cracks tend to be uniform in width with height, whereas subsidence cracks widen with

Practical Guide to Diagnosing Structural Movement in Buildings, Second Edition. Malcolm Holland.
© 2023 John Wiley & Sons Ltd. Published 2023 by John Wiley & Sons Ltd.

height due to the rotational effect. In reality expansion cracks are also a little wider at the top than the bottom because of the effect of weight. The expansion force created by heat is able to move the wall more at the top where it is lightly loaded. At the bottom there is more weight and therefore more friction, which increases the resistance to movement. The difference is, however, likely to be small and much smaller than the difference caused by rotation.

When looking at an elevation of a building it is necessary to imagine how the loads flow down and through the walls. This is like a stream or streams flowing down a hillside and becoming a river.

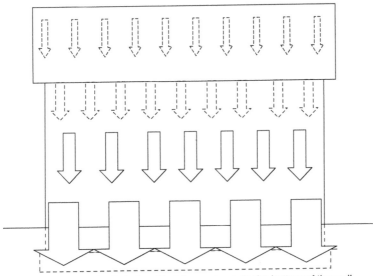

Loads evenly distributed and building up towards the bottom of the wall.

Figure 1.6.1 Simple load path.

If a building was built without any openings, in a symmetrical manner, then the loads would be evenly distributed through the roof and walls as shown in Figure 1.6.1. The loads would also build up as they travelled down through the structure. The foundations would then evenly distribute the load to the ground.

Buildings are not built symmetrically or uniformly, and there are also openings in the walls for windows, etc. Where openings are built into a wall, it is necessary to carry the weight from above, around the opening. This is achieved by beams (known as lintels) or by arches. Imagine this to be like putting a dam across the stream or putting a log or plank partly across the stream to block its path. In a real stream the water would push against the log or plank. This is the equivalent of the load trying to bend the beam or lintel. In a stream, the water would have to flow around the obstruction. At the edges of the obstruction, the water would form eddies and be turbulent. Imagine these eddies as stress points caused by the concentration of the flow of the loads.

Beams create point load concentrations at the support points. Once the flow is past the obstruction, it will start to spread out again.

In the same way, the concentrated loads then start to spread out again through the bonding of the brickwork.

Arched openings also create load concentrations. If the arch is not a complete 180° round arch, it will also create a thrust force, pushing out sideways from the opening. The flatter the arch, the more outwards thrust is created. Again, think of this thrust as the flow of water being deflected outwards.

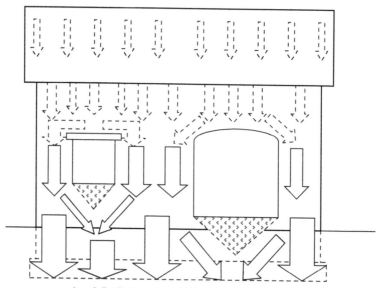

Load distribution around lintels and arches.

Figure 1.6.2 Load path around lintels and arches.

Figure 1.6.2 shows the loads passing down through an elevation where openings have been formed. The left-hand opening has a lintel above, which transfers the loads squarely around the opening and vertically around each side. The arch support over the right-hand opening creates a thrust, which pushes outwards. Imagine this as a current of water flowing out at an angle. Imagine the edge of the building as the riverbank. If the current was too strong it would overflow the bank. In a similar way, if the thrust was too strong it would 'overflow' the wall. That is, it would push the wall out.

For this reason, there must be a sufficient mass of wall either side of the arch, to provide resistance against the thrust. Again, imagine this like a flow of a river being distorted by the obstructions of the openings. Imagine the width of wall, either side of an arch, as being the distance to the bank of the river. If the bank was far enough away from the thrust, there would be time for the current of the river to turn it back into the direction of flow. If the bank was not very far away but the current was very strong, it would again be able to turn the outwards flow back into the direction of the river flow. Similarly, the thrust can be turned back by a heavy load on the piers either side of an arch.

The loads are channelled around the 'obstructions' of the openings into the channels of brickwork on each side. It is best to design buildings so that the windows are in line on each floor to avoid interrupting the flows.

Beneath the openings the load will spread out again, roughly at an angle of about 45° through the bond of the brickwork. This will leave the areas beneath the openings without any applied load. The only weight in the shaded areas is the self-weight of the brickwork. The absence of any applied load makes it very easy to move this area of the wall. It is not uncommon for hairline cracks to form through the mortar joints of the brickwork, along the junction where the unloaded brickwork meets the loaded brickwork, following the lines of the shaded areas.

This is referred to as load path cracking. An example of this, together with the associated key features, is given later in the book.

When the current of load has passed all the openings, it spreads out through the brickwork.

If the underground structure is deep enough, there will be time for the loads to spread out evenly. In the analogy of a river, think of this as the estuary, where the river slows down and widens at the mouth to the sea. The water is then distributed to the sea. In a similar way the weight of the building is distributed to the vast ocean of the ground beneath it.

If the foundations are deep enough to spread out the load, the settlement of the building will be even. That is the aim.

1.7

Movement and Orientation

Temperature and moisture variations seasonally cause expansion and contraction. There are daily changes in temperature and the diurnal gravity effect of the moon. Chemical reactions that cause movement often need water as a medium. Heat is also a catalyst of chemical reactions. Exposed elevations are therefore more likely to suffer from movement. Exposed elevations are likely to suffer a greater degree of movement.

All buildings move daily and seasonally. Movement is caused by expansion and contraction due to temperature change. This occurs mostly on the south and southwest elevations.

Movement is also caused by expansion and contraction as a result of moisture content changes. In Britain, this again occurs mostly on the south and west elevations of a building because the warm, wet winds come predominantly from these directions.

Every day a building is influenced by the gravitational pull of the moon. The gravity of the moon can lift the tides of the oceans. It also has a similar effect on the ground, but to a lesser degree.

Seasonal and daily movement is rarely significant, but it can cause other defects which aid deterioration. For example, small cracks could

Practical Guide to Diagnosing Structural Movement in Buildings, Second Edition.
Malcolm Holland.
© 2023 John Wiley & Sons Ltd. Published 2023 by John Wiley & Sons Ltd.

allow rainwater ingress, which in turn could lead to rot in structural timbers.

Movement can also be caused by chemical reactions including corrosion. As chemical reactions use moisture as a medium and heat as a catalyst, then again, any reaction is likely to be greatest on south- and west-facing elevations of buildings in Britain.

Overall, as a general rule of thumb, building defects of all types are much more common or much more severe on the elevations which are exposed to the sun, wind and rain. In Britain this is predominantly the south- and west-facing elevations.

1.8

Summary of First Principles and Process

> *By following the principles explained in this chapter most cracks can be diagnosed relatively quickly and with a reasonable degree of confidence. It is essential that the process is followed; even where at first sight the cause and effect might seem obvious. Going through the process methodically will help avoid jumping to conclusions.*
>
> *There will always be a few cracks and causes of movement that cannot be diagnosed from a single visual inspection. In such cases, further investigation will be required. This might involve opening up part of the construction, excavating trial pits to inspect the foundations, taking samples or monitoring over a period of time.*

The absolute fundamental principle is that cracking is caused by tension in the material.

The first stage of diagnosis involves sketching the building and the crack pattern in simple line form, just as in the diagrams shown in the examples given already. Then sketch in imaginary lines of tension at right angles to the cracks. The imaginary lines of tension will point in the direction of the movement, which is usually where the cause of the movement and defect actually is. In most cases the upward arrow can be ignored. Once the risk of upward movement has been considered and ruled out, look to the other direction. This is usually down, with gravity.

Practical Guide to Diagnosing Structural Movement in Buildings, Second Edition. Malcolm Holland.
© 2023 John Wiley & Sons Ltd. Published 2023 by John Wiley & Sons Ltd.

In most cases this process will immediately point to the cause of the defect which will be obvious.

If the overall crack pattern is a little ambiguous, look at the crack itself. Also take into account the factors that can distort the shape. Rotational movement and short cuts through weak routes are the most significant factors. Load distribution is also a factor, but to a lesser degree.

By looking at the actual displacement of the crack, and by adjusting the angle of the imaginary lines of tension to take these distorting factors into account, the diagnosis success rate can be improved further.

With a reasonable knowledge of building construction and with reference to the 'swatches' and key features of types of movement described in Part 2 and Part 3 of this book, there will be very few cracks that cannot be diagnosed relatively quickly.

There are, however, always some defects that cannot be diagnosed immediately. There will always be a few where the cause is unusual. In some cases, a combination of factors can make a conclusive diagnosis difficult, or impossible.

Diagnosing cracks is not always easy. The movement in a building may have to be monitored for a time before a diagnosis can be made. Some uncovering work may be required, for example, taking out bricks from a wall or excavating trial pits to inspect foundations underground. This may be inconvenient but is sometimes unavoidable.

Always go through the process from start to finish. Do not jump to a conclusion and then find the evidence to fit.

In my experience, many people start from the point of assuming foundation movement, when there are so many more likely causes of cracking than this. With this in mind, it is not an unreasonable practice to consider all other causes, before looking at whether it is potentially foundation movement. Until one gains a reasonable amount of experience my advice would be to consider and exhaust all other possible causes first. Only when all other causes have been considered and discounted, move on to whether it could be foundation movement.

Before moving on to Part 2 of the book, please look at the example below (Figure 1.8.1).

The elevation of this building shows a stepped crack approximately 0.5 mm wide running diagonally between the first-floor window and the patio door opening below. To the left there is a tree within influencing distance. It would be possible to jump to the conclusion that the crack was caused by the tree, which in this example would not be correct.

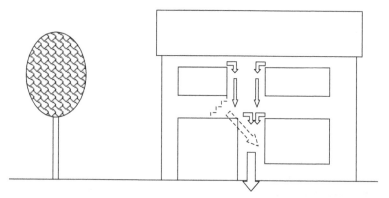

Figure 1.8.1   Arrows of tension point away from the tree.

By going through the process of sketching the elevation and the crack, and then applying an imaginary arrow of tension, it is possible to see that the tension line points down to the right, away from the tree. The movement is down to the right of the patio door opening not down to the left.

The solid arrows on the diagram show the route of the load, around and down the side of the window. The load path goes around the right-hand side of the patio door opening. The load becomes very concentrated in this narrow section of brickwork, at the right-hand side of the patio door. The load concentration here has probably resulted in long-term settlement in this area of the building over a number of years or decades. The triangle of brickwork under the first-floor left-hand window is virtually unloaded, other than its self-weight. The crack has formed at the junction of the unloaded brickwork and the heavily loaded brickwork.

The crack is therefore due to some long-term creep settlement. The movement has exploited the difference in weight distribution in the brickwork and caused a crack along this line. The movement would not be progressive to any significant amount, but if it is re-pointed, it is likely to crack again due to normal seasonal movement.

This demonstrates the importance of not jumping to conclusions. By using this methodology, a logical step-by-step process can be demonstrated. In addition, a more reliable and accurate diagnosis can be made.

Part 2

Cracks in Buildings Not Related to Foundations

Expansion Cracking

Expansion and contraction in building materials can result in cracking. Cracks are generally even in width, vertical in direction and non-progressive. In the Northern Hemisphere, south-facing elevations are more likely to suffer from cracking as they face the sun. Concrete and calcium silicate bricks have a higher co-efficient of expansion than clay bricks and are likely to suffer from expansion cracking to a greater degree. Concrete roofs can also cause horizontal cracking in supporting walls as the concrete deck expands and contracts.

Key features

1. Straight. Vertical crack.
2. Crack equal width with all displacement horizontal.
3. Any cracking on bed joint is hairline from sliding action.
4. Over sail at damp proof course on very long elevations.
5. Gaps at ends of lintels, with cracking in brickwork around bearing point.
6. No cracking below damp proof course level.
7. Usually associated with modern walls, having cement-based mortar.

Practical Guide to Diagnosing Structural Movement in Buildings, Second Edition.
Malcolm Holland.
© 2023 John Wiley & Sons Ltd. Published 2023 by John Wiley & Sons Ltd.

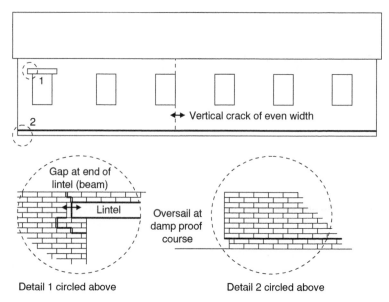

Figure 2.1.1   Expansion cracking.

Expansion cracking is one of the most common forms of cracking.

The width of the crack will depend on the co efficient of thermal expansion of the material and the temperature range. It is therefore usual for most cracking to be on the south or southwest facing elevations, exposed to the direct sunlight (in the Northern Hemisphere). In the Southern Hemisphere the cracking would be most likely on Northerly elevations.

The degree of cracking, with clay bricks, is commonly hairline up to 1–3 mm, but could be more on long elevations. With calcium silicate bricks or concrete bricks/blocks, the cracks may be larger.

The displacement is horizontal. The resultant crack, at right angles to the displacement, is therefore vertical. The crack will be even in width with height, allowing for perhaps some very minor widening at the top, due to weight distribution in the wall (See Figures 2.1.4 and 2.1.5). The absence of weight at the top of a wall makes it easier to move than it is at the base of the wall, where the weight is greatest. If the crack strays off vertical, because of weak routes through window openings, the horizontal parts of the crack will be hairline. This is because the horizontal cracks have simply been caused by the bricks sliding over each other, not because of vertical displacement (see Figure 2.1.7).

The brickwork is restrained below ground level by the soil around it. The ground also acts as a heat sink. Consequently, brickwork below damp proof course level will not expand and contract to any significant degree. As a result, there will be no cracking below damp proof course level. The damp proof course will act as a slip membrane. The brickwork above the damp proof course is therefore free to expand and over sail the damp proof course (see Detail 2, in Figure 2.1.1).

The magnitude of cracking can be more pronounced if there is a return in the elevation of a wall.

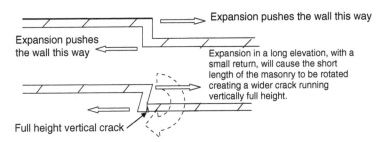

Figure 2.1.2 Expansion crack in a stepped elevation.

Expansion in the long elevations will have the effect of rotating the short return length (Figure 2.1.2). This creates a full height vertical crack, positioned at the wall's thickness from the corner of the return (see Figures 2.1.9 and 2.1.10). This rotational effect tends to make the crack wider than would otherwise be the case.

Where long concrete lintels or steel lintels are built into walls, there are likely to be gaps created at the ends, together with cracking below support points. The beams act like battering rams, pushing the walls out at either end (see Figures 2.1.6 and 2.1.8). Cracks are formed beneath the bearing points, when the brickwork is dragged horizontally by the expansion of the beams (see Detail 1, in Figure 2.1.1. See also Figures 2.1.11 and 2.1.12).

Concrete flat roof decks will also expand and contract significantly when exposed to the full heat of the sun. When a concrete roof expands, it drags the brickwork below the roof with it. This creates a horizontal crack, usually a brick course or two below the level of the roof. The crack is hairline. The expansion can also cause over sailing of the brickwork at the edge (see Detail 1, in Figure 2.1.3).

Concrete flat roofs are commonly found on single-storey outbuildings, attached to post–World War Two local authority houses, up until about the mid-1960s.

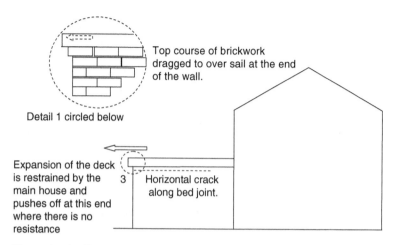

Top course of brickwork
dragged to over sail at the end
of the wall.

Detail 1 circled below

Expansion of the deck
is restrained by the   3
main house and
pushes off at this end
where there is no
resistance

Horizontal crack
along bed joint.

Figure 2.1.3   Expansion cracking related to a concrete roof.

To prevent cracks in clay brickwork, an expansion joint is required about every 10–12 m, depending on the brick type. With concrete or calcium silicate bricks, an expansion joint is required every 5–6 m.

Expansion cracking is most commonly found in post–World War Two properties, built with cement/sand mortars. Prior to World War Two, most mortar for bricklaying was made with lime and sand. Lime/sand mortar is softer and less brittle than cement/sand mortar. This explains why very long streets of terraced Victorian housing do not suffer from expansion cracking. The relatively soft, lime-based mortar is able to accommodate a little movement.

Expansion cracks are not significant in structural terms and will not get progressively worse. If, however, the cracks are pointed up, cracking will recur during the next cycle of heating and cooling.

On exposed elevations, expansion cracking can create a risk of rainwater penetration, particularly if the mortar is a cement mix, and the bricks are particularly hard. Hard bricks are not very porous, and as a result of this, the mortar does not stick to them very well during construction. Cement-rich mortar, as well as being hard, is also quite brittle. Constant repeated expansion movement breaks the bond between the bricks and mortar. This creates invisible capillary paths through the wall. As the bricks and mortar are not very porous, rainwater runs down the face of the wall, rather than being absorbed into the outer surface.

When running water meets a crack, it is drawn in by capillary action. Even where movement associated with expansion is not significant in structural terms, there can be other reasons why it is detrimental to the performance of a building, which may call for remedial action. In most cases however, as the cracks will not become progressively larger, no remedial action is usually required. If the cracks are large enough, and sufficiently exposed to allow water penetration, then they could be sealed with flexible mastic. Filling cracks in the face of a wall with flexible mastic is, however, often not acceptable from an aesthetic point of view. If the cracks are minor, it is often more desirable to leave them. For moderate width cracks of a millimetre or two, re-pointing with a soft mortar mix may allow just enough movement to avoid cracking and might be more aesthetically acceptable.

Where the cracking is more pronounced, it may be necessary to form expansion joints in the elevations. This involves cutting through the masonry with a masonry saw, to form a slot. The slot is then filled with mastic. Special sliding expansion wall ties are manufactured to connect the masonry across the formed joint. These hold both sides of the joint together in order to provide lateral stability but allow movement across the joint for expansion and contraction.

Over recent decades, changes in Building Regulations, and the desire to conserve energy, have resulted in cavity walls being filled with insulation. When the cavity is filled with insulation, the temperature variation within the outer leaf is greatly increased, both positively and negatively. When the sun shines on the face of an insulated wall, the insulation prevents the heat from being dissipated. As a result, the temperature of the outer leaf of the wall will rise much more than it would with a non-insulated cavity. When the weather is cold, the heat from the building is held inside. It cannot escape to keep the outer leaf warm. The net result is a much wider temperature range, which increases the magnitude of movement. The age of a building, or whether a cavity walled building has been retro fitted with insulation, is therefore important.

Changes in Building Regulations in the United Kingdom mean that any house built after about 1990 will usually have full or partial cavity-fill insulation. Retro-fitting old cavity walls with insulation really began after the oil shortages of the mid-1970s. Even then, insulation was installed in only relatively small numbers. Initially it was carried out by filling the cavities with expanding formaldehyde foam. Quite commonly, evidence of the foam can be found in the roof space, extruding through gaps in the mortar joints of the gable walls. Modern cavity fill usually involves blowing loose fibrous insulation, such as fibreglass or rockwool, into the cavity. The insulation is inserted by drilling 25 mm diameter holes through the brickwork at the junction of the horizontal bed joints and vertical joints (perp joints). The holes are re-pointed afterwards. These re-pointed holes are evidence that the wall has been filled.

In older examples of cavity fill, holes of about 50 mm diameter were drilled through the face of the bricks. After the insulation was inserted, the holes were filled with a coloured mortar. This gives a rather unsightly finish which did not prove very popular. It is therefore quite rare to find this. An alternative was to cut out bricks completely. After filling the cavity with insulation, the bricks were stitched back in again. It is impossible to completely match the mortar so it should be possible to spot the replaced bricks. If there is a regular pattern of individual replaced bricks, it is probably evidence of cavity fill by this method.

Care should be taken to avoid confusing re-pointed cavity insulation holes or replaced bricks with replacement cavity wall tie repairs (see also notes on Cavity Wall Tie Corrosion).

Until recently cavity wall ties were normally found at 900 mm horizontally (every four bricks horizontally) and 450 mm vertically (every six brick courses vertically). From around 2000, walls tend to have wider cavities to accommodate more insulation, and the horizontal spacing would be around 750 mm. Cavity-fill holes or replaced bricks will be wider apart than this, commonly 1350 mm horizontally (every sixth brick horizontally) and 1350 mm vertically (every 18 brick courses vertically). In addition, with cavity fill, the holes will be drilled within 300 mm of the underside of eaves, and under windows. Where window openings disrupt the pattern, the holes may need to be closer together, usually not more than 800 mm horizontally (three and a half bricks horizontally).

Figure 2.1.4   A typical example of an expansion crack, midpoint on a garage block elevation.

Figure 2.1.5   Vertical expansion crack. The crack is slightly wider at the top where the load on the brickwork is least. The bricks are calcium silicate in this case.

Figure 2.1.6   Expansion crack between ground-floor window and first-floor window. The crack is exacerbated by the concrete lintel.

Figure 2.1.7   Close-up of expansion crack. 1–2 mm horizontal displacement and just hairline cracking along the bed joint.

Figure 2.1.8   Typical gap formed at the end of a concrete lintel by expansion. The gap is quite small in this case because the window opening is quite narrow.

Figure 2.1.9   Typical vertical crack at a staggered return. See also photo 2.1.10. The timber lattice frame in front of the brickwork is just decorative.

Figure 2.1.10   Close-up of a vertical crack at a staggered return in a brickwork elevation. The timber lattice frame in front of the brickwork is just decorative. Crack approximately 1 mm. All the displacement is horizontal.

Figure 2.1.11   Expansion of concrete roof deck. Expansion of the concrete roof deck drags the brickwork horizontally creating a crack along the bed joint.

Figure 2.1.12   Close-up view of horizontal cracking, caused by expansion of the concrete flat roof deck.

2.2

Cavity Wall Tie Corrosion

Cavity walls are usually built with metal wall ties, holding the inner and outer leaf together. The metal ties corrode with time. Cracks are horizontal and usually occur on the wall tie positions, usually about every six courses. Cracks begin at the top of the walls first. Most wall ties will last in excess of 60 years unless other accelerating factors are present such as black ash mortar, industrial pollution or exposure. In order to diagnose cavity wall tie corrosion it is first necessary to be able to tell the difference between a cavity wall and a solid wall. This is a progressive defect that will become worse with time.

Key features

1. Horizontal cracks related to the position of wall ties.
2. Cracking on exposed south and southwest elevations.
3. Cracking more likely if accelerating factors present.
4. Cracks appear at top of wall first.
5. Crack may initially be inside in gable wall.
6. Walls bulge in or out as stability is lost.

Practical Guide to Diagnosing Structural Movement in Buildings, Second Edition.
Malcolm Holland.
© 2023 John Wiley & Sons Ltd. Published 2023 by John Wiley & Sons Ltd.

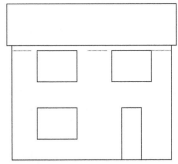

First signs of cracks just below eaves where weight on brickwork is least.

Figure 2.2.1 First evidence of tie corrosion.

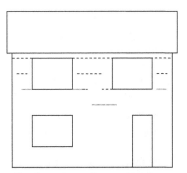

Cracks progress from eaves down horizontally on lines of ties.

Figure 2.2.2 Progression of cavity wall tie corrosion.

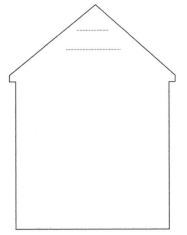

First signs first course of ties above purlin positions.

Figure 2.2.3 Evidence of wall tie corrosion in the gable wall.

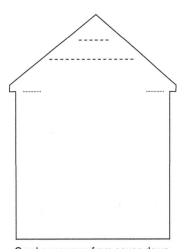

Cracks progress from eaves down horizontally on lines of ties.

Figure 2.2.4 Progression of tie corrosion in the gable walls.

Cavity walls are built with an inner and outer leaf of brickwork, with a cavity in between. The inner and outer leaves are usually connected together by iron or steel ties. The ties are built into the mortar joints, and connect the two leaves of masonry across the cavity.

With time the ties can corrode. The products of corrosion (rust) have a much larger volume than the un-corroded metal. This causes it to expand and lift up the brickwork. As the weight on the brickwork is much more at the bottom of the wall than at the top, it is easier for the corrosion to lift the brickwork at the top than at the bottom. As a result, the cracking begins at the top of the wall and increases progressively from the top to the bottom. It is likely that the defect will be detected at the early stages of development and rectified before the cracks progress all the way down the wall.

The first sign of cracking will usually be horizontal cracks just below eaves level (Figures 2.2.1 and 2.2.8) or above purlin level in gable walls (Figures 2.2.3 and 2.2.6). Cracking is also more likely in areas of brickwork that are less heavily loaded, for example, directly beneath openings.

The cracks may also be noted internally in gable walls within the roof space, where the loading is particularly light. The very first sign of cracking is sometimes internally within the roof space. This is because the ties begin to rust where they are built into the inner face of the external leaf. This wedges up the mortar joint at this point, causing a slight outward curve to the gable (Figures 2.2.5 and 2.2.10). At this early stage of corrosion the wall ties are still strong enough to transmit the

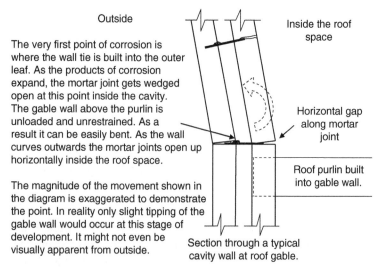

Figure 2.2.5   Internal cracking in the gable wall from tie corrosion.

effect of the curve across the cavity to the inner leaf. There is no load on the gable wall above the line of the purlins. It can be easily bent and lifted open along the joints as a result (see Figures 2.2.7 and 2.2.9).

As the defect develops, the gaps increase in size and new gaps/cracks develop lower down the walls (Figures 2.2.2 and 2.2.4). Cracks may also begin to develop internally as well. The cracks will usually occur horizontally in bands, usually six courses apart. This corresponds with the normal, or most common, spacing of wall ties in a cavity wall. It is, however, not a hard and fast rule. Ties could be spaced at any number of courses. The most common spacing is every six courses, but it could be every four, or every eight, or indeed any other number.

As the ties corrode completely they will snap. The two leaves can then separate and the wall may bulge. Usually the outer leaf will bow outwards, rather than the inner leaf bowing inwards. This is because stability is provided to the inner leaf by built in floor joists. Gable walls, however, may bow inwards, as there is often little or nothing in the roof construction to prevent this.

The age at which a building is likely to be at risk of wall tie corrosion is also influenced by a number of possible 'accelerating factors'.

Accelerating factors include:

1. Acidic black ash mortar.
2. Exposure, coastal or general.
3. Coastal areas, where salt in the atmosphere makes corrosion of iron more likely.
4. Industrial areas where rain is more acidic.

The corrosion is most likely to occur on exposed elevations, so generally it is most likely to occur on southwest elevations first.

Corrosion is a chemical reaction and heat is a catalyst for chemical reactions. This makes the south/southwest elevations more at risk, as they are warmed by the sun.

The risk is also associated with the general age of the building.

Although cavity wall construction was not unknown before World War One, it was not very common. The most common form of construction before World War One was solid 225 mm brickwork. In the relatively few pre World War One houses with cavity wall ties, the mortar is likely to be lime based. This does not provide the same degree of alkaline protection that a cement mortar would. Any cavity wall house, as old as pre–World War One, will be at risk.

Virtually all houses built after World War Two have cement-based mortar. This provides an alkaline environment, which protects the ties from corrosion. At present whilst there may be some risk in cavity walls built in the 1950s and 1960s, the vast majority of wall tie corrosion problems will be in houses built prior to World War Two, which are in excess of 60 years old.

Interwar housing could have cement-based mortar, but lime-based mortar would still be most common. Corrosion is more likely in lime-based mortar. The alkalinity of cement-based mortars will be reduced over time by carbonic acid in rain. This occurs most in industrial areas and on exposed rain-washed elevations.

When cavity wall tie corrosion is suspected, it will be necessary to confirm this by removing bricks to open up the wall, or drilling holes in the walls for borescope inspection. Removing bricks is the best option, as it allows a more thorough inspection of the condition of the ties. In practice, not many owners would welcome this and hence borescope inspection is more common.

Obviously, cavity wall tie corrosion can only be found in a cavity wall. It is therefore essential to be able to identify whether a wall is either cavity or solid construction.

Walls can be identified by their thickness and the pattern of the brickwork (the bond).

A brick is about 100–105 mm thick, nominally 102.5 mm average. In the age of building likely to be at risk, the cavity in a wall is most commonly about 50 mm wide. A cavity wall is usually made up with two leaves of brickwork with a cavity in between and plaster internally. The plaster is usually about 12–15 mm thick. The overall thickness is therefore 102.5 + 102.5 + 50 + 15 = about 270 mm in total. This can vary by about 10 mm either way allowing for building accuracy. The pattern of the brickwork is usually stretcher bond; that is, all the bricks show their long face externally with a half brick lap at each course.

Solid walls are usually 215 mm thick brickwork plus plaster; about 230 mm in total. Occasionally, they will be 325 mm thick plus plaster; about 340 mm in total. There are a number of brick bonds for solid walls but all have the small face of the brick (known as headers) visible in places. In solid walls there is a quarter brick lap at each course. The most common bonds are Flemish garden wall bond, Flemish bond, English garden wall bond and English bond.

There are some variations to this simple guide. An example would be if a cavity wall extension is built to a solid wall house. Sometimes the bond of the original solid wall is often copied in the cavity wall, for the sake of aesthetics. In order to do this, bricks have to be snapped in half and put into the outer leaf of the cavity wall with the small face of the brick facing outwards. These are called 'snap headers'. In this case the thickness of the wall will be 270 mm but the pattern or bond of the bricks will be a solid brick bond in appearance.

Over recent years cavity walls are often built with cavities wider than 50 mm to incorporate cavity wall insulation. These walls are often 300 mm thick or even wider. Plasterboard dry lining is also often used instead of plaster applied directly to the wall. Depending on how this is fixed, it will commonly add 30–40 mm to the overall thickness of a wall. Walls of this age, however, would not be at risk of tie corrosion.

In old examples of cavity wall construction, there may be a 215 mm inner leaf, a cavity of around 50 mm and then a 105 mm thick outer leaf of brickwork. With plaster this would be around 330–340 mm thick. With this type of construction snap headers may have been used in the outer leaf to achieve a traditional solid wall bond pattern,

There is also a 'rat trap' bond where a part solid and part cavity wall is achieved by laying the bricks with the face down on the bed joint of mortar. This type of wall is often found in very cheap, old, artisan housing and will not suffer from wall tie corrosion (for the simple reason that it does not have metal ties).

Brickwork is a complete subject in itself. Brick patterns and more detailed information can be found in many text books on brickwork and building construction.

There is some debate about whether cavity wall insulation might increase the risk of corrosion. The presence of insulation will hold heat inside the building. This will delay the time that it will take for an exposed wall to dry out after cold wet weather.

The outer leaf will be colder. This could increase the chance of moisture vapour condensing in the outer leaf, making it damper for longer.

During warm weather, the temperature of the outer leaf will be higher. As heat and moisture are both factors that are likely to increase the rate of corrosion, cavity wall insulation might increase the risk. As yet there is no reliable evidence for this.

Figure 2.2.6   First external signs of wall tie corrosion in a rendered gable wall in a 1930 house.

Figure 2.2.7   Horizontal crack and bowing of a gable wall inside the roof space of a 1930 house. The crack is approximately 2 mm wide and was the only evidence of tie corrosion visible either internally or externally.

Figure 2.2.8 Cavity wall tie corrosion developing down the wall, in a house built circa 1900. The cracks are spaced six courses of brickwork apart, corresponding with the spacing of the ties (photograph courtesy of Caroline Legg).

Figure 2.2.9 Horizontal crack and bowing gable wall in a 1908 house. Note stretcher bond internally. Externally the wall has 'snap' headers to look like a solid wall (see Figure 2.2.10).

Figure 2.2.10   Inward bowing to south west gable wall in a 1908 house.
Note on the front elevation, the 'snap' headers in external face of the wall to
give the appearance of a solid wall.

2.3

Corrosion of Metal Built into Walls

Windows and door frames are often held in place with iron or steel fixings. Lintels in cavity walls are also often steel. Iron and steel corrodes with time. The metal expands and causes localized cracking.

Steel or iron built into walls inevitably corrodes with time. When such metals corrode there is an increase in volume from the products of corrosion. This volume increase will push apart material, causing localized cracking (Figure 2.3.1). Commonly iron lintels are found in interwar or early post–World War Two buildings (see Detail 1 in Figure 2.3.1). Iron cramps or brackets are commonly used to fix window and door frames within brickwork openings (see Detail 2 in Figure 2.3.1).

Similar cracking is also likely to occur around iron air bricks commonly built into external walls, to vent sub-floor voids, wall cavities and bedrooms. Corroding air bricks will lift brick joints up. If the air bricks are near the corners of the walls the corrosion expansion can push bricks out sidewards (see Figures 2.3.2 and 2.3.3). Iron air bricks are common in houses with suspended floors up to World War Two. Air bricks in external walls are commonly found in houses of the interwar period.

Practical Guide to Diagnosing Structural Movement in Buildings, Second Edition.
Malcolm Holland.
© 2023 John Wiley & Sons Ltd. Published 2023 by John Wiley & Sons Ltd.

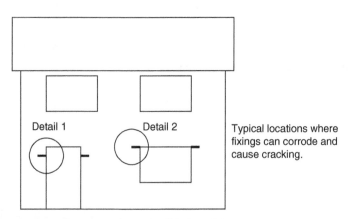

Detail 1   Detail 2   Typical locations where
                                    fixings can corrode and
                                    cause cracking.

Figure 2.3.1   Corrosion of metal built into walls.

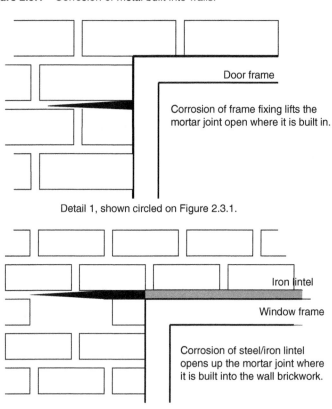

Door frame

Corrosion of frame fixing lifts the
mortar joint open where it is built in.

Detail 1, shown circled on Figure 2.3.1.

Iron lintel

Window frame

Corrosion of steel/iron lintel
opens up the mortar joint where
it is built into the wall brickwork.

Detail 2, shown circled on Figure 2.3.1.

Figure 2.3.2   Corrosion of the iron air brick in a 1930 house. Corrosion of the air brick at the top left-hand corner has pushed out the corner brick.

Figure 2.3.3 Close-up of the corner brick pushed out of the wall by expansion from the corroding air brick.

2.4

Vibration of Built-in Fixings

> *Door frames are held in place with metal fixings. Repeated use of the door causes vibration around the fixing. This can cause localized cracking. It is not a serious defect.*

Door frames are commonly fixed into walls using metal brackets, (known as 'cramps' in the building industry) which are set into the mortar joints of the masonry. Every time a door is opened and closed there is a little vibration at the fixing point. If the door is slammed or pulled open quickly to the full extent of the hinges, the tug at the fixing will be greater. In most cases this has no effect because there are numerous fixings, evenly positioned around the opening, and the masonry is strong enough to resist the force. There is however, a very common situation where cracking will commonly be seen, which is caused by vibration around the fixing.

It is very common for the external kitchen door unit to be combined with a window. The load travels from the lintel (beam) over the opening, and down the side of the combined window/door opening. When the load reaches the bottom of the window opening, it spreads out again, through the bond of the brickwork. This leaves a small triangle of brickwork directly beneath the window, and to the side of the door, completely unloaded. The mid-height door cramp, holding the door in place, is usually fixed into this unloaded triangle of brickwork.

Practical Guide to Diagnosing Structural Movement in Buildings, Second Edition.
Malcolm Holland.
© 2023 John Wiley & Sons Ltd. Published 2023 by John Wiley & Sons Ltd.

Every time the door is opened and closed it vibrates this brickwork (Figure 2.4.1). There is no weight on the brickwork to hold it in place, and so the repeated vibration eventually loosens it. A crack will commonly occur on the junction line between the unloaded area and the loaded area (see Figures 2.4.1 and 2.4.2). This cracking is usually hairline to 1 mm. It is not a progressive structural fault. With time however, it can loosen the brickwork to the extent that the door frame itself becomes insecure. If the brickwork loosens to the extent that it has to be repaired, it should be cut out beyond the line of the load path and rebuilt locally. Unless it is stitched back in with an epoxy mortar and with reinforcing mesh in the joints, it is likely to loosen again with time.

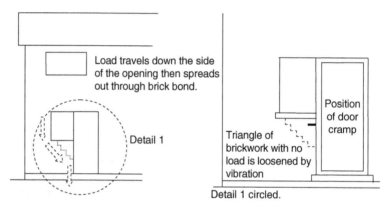

Figure 2.4.1   Vibration of door frame fixing.

Figure 2.4.2   Re-pointed crack caused by vibration of the door fixing.

Roof Spread

If a roof structure is not properly triangulated, roof spread can occur. The roof can push the walls out of vertical at the top. The walls may also bow. It is often related to poor design or alterations, such as re-roofing with a heavier material than the original roof covering. Tie bars are often used to restrain walls. Movement can be progressive and serious.

Key features

1. Poorly triangulated roofs.
2. Heavy roof coverings.
3. Walls pushed out of vertical.
4. Separation of inner walls from external walls.
5. Tapering cracks usually wider at top.
6. Displacement is greatest at the mid span of the purlins.
7. Displacement increases between restraining walls.

Practical Guide to Diagnosing Structural Movement in Buildings, Second Edition. Malcolm Holland.
© 2023 John Wiley & Sons Ltd. Published 2023 by John Wiley & Sons Ltd.

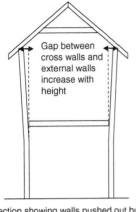

Section showing walls pushed out but restrained to first floor level by floors

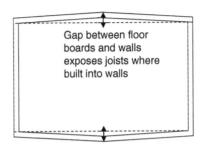

Plan showing walls bowed out where least restrained.

Figure 2.5.1 Roof spread.

Roof spread is caused by a poorly triangulated roof. If a roof structure is poorly triangulated it allows the rafter feet to spread and push out the top of the wall. The pattern of movement can vary a little depending on several factors, but Figure 2.5.1 illustrates the most common features.

The roof pushes the wall out at the top. Cracking occurs between internal walls and the external walls. The cracking is wider at the top than the bottom. The internal cracking is often covered by decoration over the life time of the building or in an effort to conceal it. If they have been wall papered over, the wallpaper will stretch to some degree before it rips.

Look for signs of buckling in the wallpaper at corners. Also look for signs of tearing in the wallpaper.

If the wallpaper has torn, this is caused by tension (being pulled apart). If the wallpaper is buckled, this is caused by it being pushed together (compression). If wallpaper tears at right angles to the movement (the angles of tension) then it compresses in the same direction as the movement. Please refer to Part 1, Chapter 1.1 where compression and tension, as a result of movement, is explained.

The magnitude of this push (thrust) will depend largely on the weight of the roof-covering material.

A well-designed roof is constructed to form a 'closed triangle'. The rafters are connected at their feet with a horizontal timber, usually forming the ceiling (see Figure 2.5.2). The ceiling joists should be nailed to the rafter feet, in order to provide a restraining connection. This ties in the rafter feet and is referred to as the 'ceiling tie'.

Without a ceiling joist to tie in the rafter feet, there will be a tendency to thrust outwards. If the roof covering is heavy enough, and the walls do not have sufficient mass to resist the thrust, roof spread will occur.

Figure 2.5.2 is a fully triangulated roof and Figure 2.5.3 has no triangulation at all. In practice many roofs are designed and built somewhere in between and examples are shown in Figures 2.5.4 and 2.5.5.

Roofs of a hipped design also suffer from roof spread sometimes. This is because the ceiling joists normally run front to rear, leaving the flank wall and the hip rafters poorly restrained. To provide restraint to the rafter feet, along the flank wall, horizontal timbers should be run across the ceiling rafters at right angles (Figure 2.5.6). These timbers are known as binders. These should be nailed to several of the rafter feet along the flank wall to hold them in. If the binders are not put in during construction the roof can spread.

Hip ties (also known as dragon ties) are also recommended across the corner of the walls at the foot of the hip rafter. Hip ties, however, are rarely seen in any but the best-quality hipped roofs. Hipped roofs are most common in interwar housing, particularly semi-detached designs.

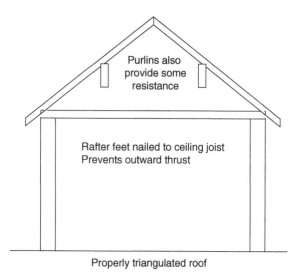

Purlins also provide some resistance

Rafter feet nailed to ceiling joist Prevents outward thrust

Properly triangulated roof

Figure 2.5.2   Triangulated roof structure.

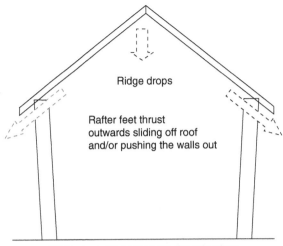

Ridge drops

Rafter feet thrust
outwards sliding off roof
and/or pushing the walls out

Without a ceiling joist to tie the rafter feet
together, the rafters push the walls out.

Figure 2.5.3   Roof spread.

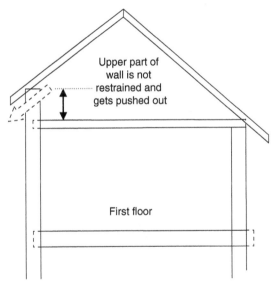

Upper part of
wall is not
restrained and
gets pushed out

First floor

Non symmetrical roof leaves front wall
poorly restrained

Figure 2.5.4   Non-symmetrical roof spread.

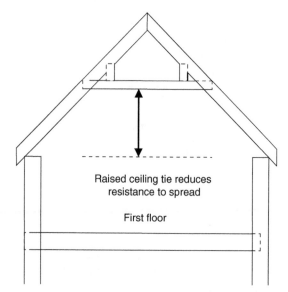

Raised ceiling tie reduces
resistance to spread

First floor

'Raised Tie' roof design offers limited
restraint

Figure 2.5.5   Raised tie roof.

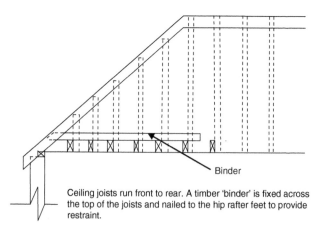

Binder

Ceiling joists run front to rear. A timber 'binder' is fixed across
the top of the joists and nailed to the hip rafter feet to provide
restraint.

Figure 2.5.6   Hip roof construction and binders.

Completely unrestrained roofs are relatively uncommon. Raised tie roofs are relatively common. They are often found in old cottages. Raised tie roofs are also a particular design feature, commonly found in 'Victorian Gothic'-style architecture.

With raised tie designs, the amount of resistance to spread decreases as the tie is raised further up the roof. Ideally, as a rule of thumb, the tie should not be more than a third of the way up the total height of the roof. It is commonly found about half way up the height of the roof in Victorian Gothic design. It was often possible to get away with this, because a very lightweight roof covering of slates was usually used. When used with a very lightweight slate covering, the roof design was able to resist spread. Problems often arise later, when the original slate roof covering reaches the end of its performance life, and is replaced by a much heavier (and cheaper) tile covering. Slate roof coverings commonly last about 100 years before the nails rust away.

The mass of the wall is also a factor. Heavy stone walls, or walls containing buttresses, have more mass to resist the roof spread force, by friction and gravity. It is also not uncommon to see single-storey porches, on the front elevation of old cottages, acting as a buttress to the walls. This is often by chance rather than by design.

The walls of a building are more rigid at the corners and this stiffens them. Bulging in the walls is likely to be more pronounced at the mid span of the wall, where there is the least restraint.

Floor joists built into the walls will also provide some restraint, by friction. If the floor joists run parallel to the wall, spread is more likely to occur, and the magnitude is likely to be greater.

When a roof spreads out, the height of the ridge comes down. Purlins running at right angles to the rafters prevent, or limit, this by holding the rafters up. In older buildings, the purlins are not usually big enough or designed in such a way as to completely hold up the roof. The purlins will usually sag in the middle, allowing more spread in relation to the centre of the purlin span, than at the supports. Purlins set at right angles to the rafters provide less resistance than purlins set vertically.

The most common symptom of roof spread is that the walls have been pushed out of vertical. This can be measured with a spirit level or by hanging a plumb bob from a first-floor window. Hanging a plumb bob gives a better overall impression of just how the wall has moved out of vertical and/or bowed.

Bowing in the wall is a common feature, depending on which parts of the wall and roof are most restrained. The influencing factors for this have been discussed above.

Roof spread in post-war houses is relatively rare. Modern truss rafter roofs are virtually universal after around 1970. This type of roof structure is, by its nature, triangulated and therefore not vulnerable to roof spread. Between 1950 and 1970, roofs were commonly of a partial truss design incorporating TRADA trusses (Timber Research and Development Association). Again this type of roof is triangulated and not prone to spread.

Sometimes a roof will be part trussed and part carpenter's roof, built on site. Even today there will be some occasions where a carpenter's roof is built, but it will be relatively rare.

On rare occasions, a particular type of defect may be seen, where a carpenter's roof, or part carpenter's roof, has been used, and not built properly. Figure 2.5.7 shows the effect where roof spread has tipped over the top course of blockwork into the cavity of a cavity wall. The lintels in a modern building are likely to be box steel. This provides a weak horizontal line at window head height, about one block down from the ceiling.

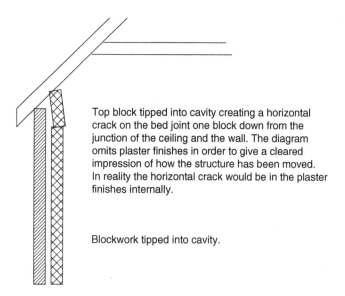

Top block tipped into cavity creating a horizontal crack on the bed joint one block down from the junction of the ceiling and the wall. The diagram omits plaster finishes in order to give a cleared impression of how the structure has been moved. In reality the horizontal crack would be in the plaster finishes internally.

Blockwork tipped into cavity.

Figure 2.5.7   Roof spread tipping blockwork into the cavity.

The particular feature of this type of spread is horizontal cracking on the lintel bed joint, and/or the bed joint of the blockwork, one block below first-floor ceiling level. If the cavity has taken up all the movement, there will be no evidence of any movement externally.

2.6

Springing from Deflected Beams

Overloaded and sagging roof purlins or beams can create an outwards thrust on walls. The thrust is resisted by the mass of the masonry. Gable walls are often lightly loaded and vulnerable to distortion. It is often related to poor design or alterations, such as re-roofing with a heavier material than the original roof covering. Movement is usually restricted to the gable walls, where it can cause instability.

Key features

1. Distorted walls (usually gable walls to a roof).
2. Commonly half brick thick or part half brick thick gables.
3. Deflected timber beams (usually purlins in a roof).
4. Commonly found in low spec artisan housing of the Georgian and Victorian period.

Practical Guide to Diagnosing Structural Movement in Buildings, Second Edition.
Malcolm Holland.
© 2023 John Wiley & Sons Ltd. Published 2023 by John Wiley & Sons Ltd.

Gable wall 110 mm (4 inch) single skin brickwork
with 215 mm (9 inch) piers for purlins

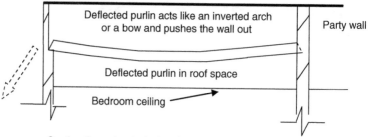

Section through a typical roof space showing deflected purlin
(rafters omitted for clarity).

Figure 2.6.1   Deflection of roof purlin.

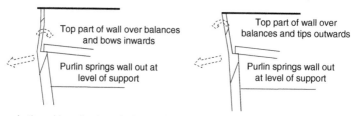

As the gable wall gets pushed out at the purlin support point, it may over balance. It may tip inwards and bow: or alternatively it may over balance externally and lean outwards. The movement depends on whether it was slightly out of vertical in the first place, or to what extent it is connected at the purlin and ridge. In the diagrams the position of the rafters has been omitted for clarity.

Figure 2.6.2   Gable wall bows inwards.

Figure 2.6.3   Gable wall bows outwards.

This defect is most commonly found in gable walls of houses. Springing caused by deflection of a beam can occur anywhere where there is a deflected beam, and a lack of restraint to resist the springing effect.

The gable wall of a house is often quite poorly restrained. It only became a requirement to connect the gable wall to the roof structure, using galvanized metal straps, from around 1980 onwards. Prior to this, the only lateral restraint would be provided by the purlins of the roof, being built into the gable wall. The purlins are usually built in at mid-height, leaving the top half of the gable wall completely unrestrained.

Under normal circumstances, where there is a reasonably sized purlin, and with a 215 mm (9 inch) thick brick gable wall (or 280 mm cavity wall in modern houses), the gable is likely to remain reasonably plumb and straight. In older buildings, or poor-quality buildings, it is not uncommon to find that the gable wall is not a full brick thick. Sometimes the gable may be built a full brick thick (215 mm), up to the purlin line, and then half brick (110 mm) above. The gable wall may just be a half brick thickness for its full height with 215 mm piers at the purlin supports points. The worst case of all is that the gable is only half brick thick for the full height, with the purlins being supported on corbelled brickwork projections.

When the gable brickwork is of poor-quality construction, it is more likely to become unstable, as a result of outwards thrust from the purlin.

If the roof purlin is too small, it will deflect and sag under load and with time. As the purlin sags, it becomes bowed and acts like an inverted arch. Rotation is created at the support point and the vertical load is turned to create an outwards thrust (see Figure 2.6.1).

The wall is pushed out at the purlin support point. As it is pushed out, the top half of the wall may overbalance either inwards or outwards. If the purlins are built into the wall, then it is more likely to overbalance inwards (see Figure 2.6.2). The very top point of the wall is often prevented from tipping in by the ridge board and the result is an inwards bow in the gable wall.

If the purlin connection is not tightly built in, the wall is less likely to tip inwards and may just lean out or bow out (see Figure 2.6.3).

To some degree, whether it bows inwards, or outwards may simply just depend on whether the wall was built completely vertical to start with.

In many old houses the walls will have been built to a poor standard and the timbers may be small, but even so, no movement occurs. This is often due to the use of a very light roof covering such as traditional Welsh slates. Quite commonly, the defect is caused by replacing an old slate roof covering, which has reached the end of its performance life, with a heavier roof covering such as concrete tiles. Modern concrete interlocking tiles are likely to be approximately one and a half to twice as heavy as traditional slates. Concrete plain tiles are likely to approximately two to three times as heavy as slates. This additional weight causes the purlin to sag and creates an outwards thrust on the gable wall.

If the movement is minor, then the problem can be resolved by strengthening the purlin and thereby removing the outwards thrust. Additionally, lateral restraint could be provided by straps or tie bars, fixed between the roof structure and the walls.

In the worst cases the gable wall may need to be taken down and rebuilt.

2.7

Lack of Lateral Stability

Masonry walls are thin and slender. Stability and rigidity are achieved traditionally by a number of design features such as small room sizes, buttresses, chimney breasts, corners and floor connections to walls. Lack of lateral restraint is often associated with poor design or alterations, such as the removal of internal walls. Movement can be progressive and serious.

Key features

1. Removal of cross walls and chimney breasts.
2. Inadequate returns at corners of walls.
3. Absence of restraint from intermediate floors.
4. Leaning walls.
5. Bulging walls.
6. Separation of internal walls from external walls.

Firstly, we have to understand what is meant by lateral stability, before a lack of it can be appreciated (Figure 2.7.1).

Imagine a free-standing wall. If a free-standing wall was pushed sideways, it could quite easily be overturned. If the wall was pushed down its length, it would be virtually impossible to overturn. If the wall was loaded directly down its centre axis (pure compression), it would not be crushed.

Practical Guide to Diagnosing Structural Movement in Buildings, Second Edition.
Malcolm Holland.
© 2023 John Wiley & Sons Ltd. Published 2023 by John Wiley & Sons Ltd.

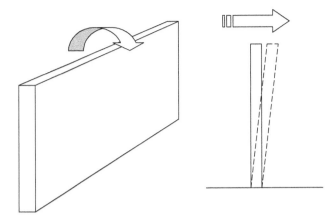

Free standing wall easily pushed over at right angles

Figure 2.7.1 Lateral instability in a free-standing wall.

A wall like this might be strong in pure compression and in one dimension down its length. In practice, however, it would bow or buckle. It would fail prematurely at right angles to its length, due to this lateral instability.

Buildings are usually built as rectangles and not as free-standing walls. The corners of a building and the intersections of internal walls provide stability at right angles to the main length of any wall.

Imagine a box containing 12 wine bottles with cardboard dividers inside. The cardboard divides the box up into smaller cells, at right angles to each other. With the bottles removed, but with the cardboard dividers still in place, the whole box will be rigid. If the cardboard dividers are removed, the box will become less stable and could be easily distorted or collapsed completely.

Long, straight walls without any form of internal division or buttresses are weak laterally. Straight walls with little or no returns at the corners are also weak.

Historically, a number of principles have been incorporated into the design of masonry buildings to keep them rigid:

1. Small rooms.
2. Chimney breasts.
3. Floor connections to walls.
4. Roof connections to walls.

Room sizes were generally kept quite small, with commonly 3–4 m as a maximum span. This created a cellular structure of small rigid 'boxes', making up the overall 'box' of the building.

Chimney breasts built into the walls create large buttresses of brickwork which add rigidity to the structure.

Floor and ceiling joists, built into the walls at right angles to the brickwork, restrain the walls. Similarly, a well-triangulated roof also provides restraint to the walls.

In more modern buildings, there is a demand for larger room sizes or even open plan design. Modern buildings rarely incorporate more than one chimney stack. Quite commonly there is no chimney stack at all.

To overcome this reduction in lateral stability within the walls, modern buildings rely more on the floors and roof to provide rigidity. The floors and roof are connected to the walls. Where floor joists are built into the supporting walls at the bearing points, this serves to 'tie' in the wall at this point and stiffens it. In larger, commercial style buildings, concrete intermediate floors are used to impart greater rigidity. These floors are often reinforced to span two ways, like a grid, to provide rigidity all round the 'box' of the building or room.

In most buildings the floors only span in one direction, either front to back or side to side. Traditionally, where the floors are parallel to the walls, they did not provide any restraint.

From around 1976 onwards, it became normal practice to mechanically connect the floors and the roof structure to the external walls.

The walls are connected to the floors and roof by 'L'-shaped galvanized mild steel straps usually 30 mm × 5 mm section, and as long as is required to span across the adjacent timbers (Figure 2.7.2). Although the straps are referred to as 'L' shaped they are in fact used horizontally and the hook of the 'L' is much shorter than the leg of the 'L', that is like '¬' in practice.

Historically, most buildings were built in a stable manner, and this is still true today. The few that were not built in a stable manner would probably have failed soon after construction or at least quickly displayed evidence of impending failure that required remedial work. The evidence of repair work can sometimes be seen on some older buildings, where tie bars have been inserted to hold them together. The evidence is often a metal plate on the external face of the building, commonly round in shape or an 'X' or an 'S' shape.

Some old buildings (and modern ones) do have parts of the structure that are poorly restrained, for example, where a staircase opening is next to a wall. The formation of the opening for the staircase makes it impossible to tie the floor to the wall, where the opening in the floor is located.

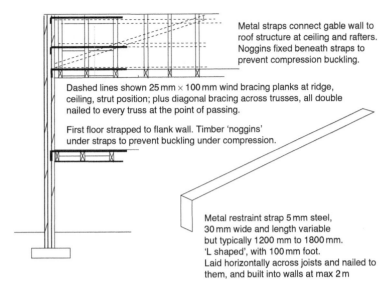

Metal straps connect gable wall to roof structure at ceiling and rafters. Noggins fixed beneath straps to prevent compression buckling.

Dashed lines shown 25 mm × 100 mm wind bracing planks at ridge, ceiling, strut position; plus diagonal bracing across trusses, all double nailed to every truss at the point of passing.

First floor strapped to flank wall. Timber 'noggins' under straps to prevent buckling under compression.

Metal restraint strap 5 mm steel, 30 mm wide and length variable but typically 1200 mm to 1800 mm. 'L shaped', with 100 mm foot. Laid horizontally across joists and nailed to them, and built into walls at max 2 m

Figure 2.7.2 Lateral stability by restraint straps in modern construction.

In pre–World War One buildings, brick arches are commonly built too close to the end of the elevation. The thrust created by the arch cannot always be held in by the brickwork. This specific type of movement 'arch thrust' is discussed individually later in the text.

In modern buildings, something as common as poorly filled 'perp' joints can result in a significant reduction in the lateral stability of an elevation (perp joints are the vertical mortar joints in brickwork, 'perpendicular' to the bed joints).

The perp joints should be completely full of mortar to firmly wedge the bricks in place within the wall. If the perp joints are not completely filled, the bricks are more able to twist. As a result, the wall becomes more flexible, allowing it to distort.

The most common cause of lateral instability is ill-advised alterations. Over the lifetime of a building, it is quite common for alterations to be carried out. Internal walls might be removed, for example. Even if the walls are not load bearing, they are likely to provide a restraint function to the walls running at right angles to them. Beware of large open-plan floor layouts, particularly in houses built before 1950.

When central heating superseded the need for open fires in each room, chimney flues became redundant. It was not uncommon for chimney breasts to be removed to enlarge rooms. In many cases the chimney breasts are removed from the rooms, but the stack within the roof and above the roof line is left in place. This not only reduces the lateral stability of the wall but also creates an eccentric load at the top. The eccentricity can overbalance the wall and increase the risk of it bending out of vertical.

Every case has to be judged individually because the alterations will vary in each case. As a result, there is no 'textbook' crack pattern for this type of movement. There are, however, a number of common features. These include the wall moving out of vertical or bowing. Bowing can occur vertically, horizontally or in both planes, depending on which parts are unrestrained.

Lateral instability can be very serious. Once a wall of a building has moved out of its intended vertical position, it becomes less able to resist the forces applied to it. In effect, lateral instability creates the risk of more lateral instability, accelerating to collapse.

The term, 'accelerating to collapse', is perhaps an overdramatic use of language in the context of buildings. Luckily, movement in buildings usually develops rather slowly and the symptoms are more commonly noticed, diagnosed and repaired before they become critical.

Figure 2.7.3 shows a typical example of how lateral instability often occurs. This example is an end terrace Victorian house, where the chimney breasts and main centre wall have been removed. Quite typically the floor joists in a house of this design will span front to rear and give no restraint to the flank wall. Taking out the wall and chimney breasts removes all restraint to the flank wall.

Remedial action to rectify this type of movement will vary, depending on the circumstances and degree of movement in each case. Traditionally the introduction of a tie bar through the building, from face to face, would be most common. Alternatively, a tie bar connecting the flank wall to the floor and roof could be used. Nowadays mild steel galvanized straps may be used, as an alternative to tie bars, in order to restrain the wall back against the floor or roof.

Another traditional method was to build piers externally, to buttress the walls. Raking piers (raking is the term for piers that slope) will sometimes be seen in older buildings angled back to hold the walls in.

When walls and chimney breasts have been removed, the foundations will usually remain. It is therefore often possible to reinstate walls and chimneys, provided of course that the user of the building will tolerate the change to the internal floor plan.

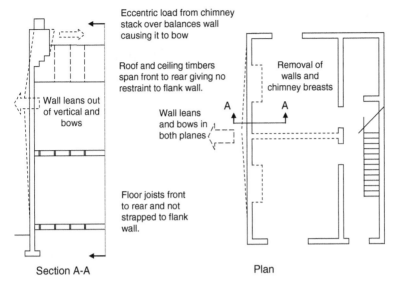

Figure 2.7.3 Lateral instability in a typical terrace house.

If the movement has developed to such a degree that none of these remedial methods can be used, then demolition and rebuilding may be the
only solution in the worst-case scenario.

Lack of Lateral Stability in Modern Gable Walls

From the late 1960s onwards, the traditional form of roof construction was largely superseded by lightweight truss rafter roof construction. The trusses are manufactured off site and erected on site. When this form of roof construction was introduced, there was no consideration given to lateral stability, at right angles to the span of the trusses. Without restraint the trusses can distort or tip over 'domino' style. A lack of connection between the gable walls and the roof trusses can also lead to instability in the gable wall. From the mid 1970s onwards, regulations were introduced to progressively require increased levels of bracing, to prevent movement.

Prior to the late 1960s, most roofs were constructed on site by carpenters, with rafters (sloping timbers) and purlins (horizontal timbers). The horizontal purlins were built into the gable walls, providing some lateral stability at right angles to the gable.

From the late 1960s, trussed rafter roof construction was introduced. Trusses are manufactured off site and delivered on a lorry. The trusses are then erected on the walls at centres of 600 mm, spanning from the external wall on one side of the building to the external wall at the opposite side. The internal walls provide no intermediate support. When this type of roof was first introduced into Britain, they were built without

Practical Guide to Diagnosing Structural Movement in Buildings, Second Edition.
Malcolm Holland.
© 2023 John Wiley & Sons Ltd. Published 2023 by John Wiley & Sons Ltd.

any form of wind bracing. They relied on the tile battens, and the weight of the tiles, to provide lateral stability at right angles to the span of the trusses. In many cases, the bracing effect provided by the tiles has proved adequate. In some cases, however, the roof trusses have distorted by bowing or 'snaking'. 'Snaking' is where the rafter section of the truss first bows one way and then the other. In some cases, the roof trusses have moved out of vertical in a 'domino style' (see Figure 2.8.1). In other cases, the trusses have remained in place, but the gable wall has become unstable and moved out of vertical, independently of the roof trusses.

The level of risk depends mostly on the degree of exposure. The risk of wind damage and overturning is greater near to the coast or on high ground, than it would be inland. There is, however, the possibility of local wind effects. For example, a narrow gap between adjacent buildings could cause local turbulence creating a suction effect. Orientation may also increase the risk of wind suction as turbulent air spills off the end of a roof.

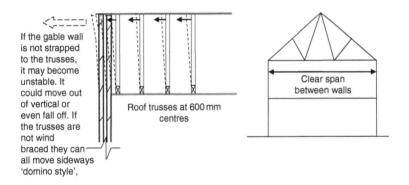

If the gable wall is not strapped to the trusses, it may become unstable. It could move out of vertical or even fall off. If the trusses are not wind braced they can all move sideways 'domino style',

Roof trusses at 600 mm centres

Clear span between walls

Section Through the Gable and Roof

Figure 2.8.1 Lateral instability in a truss rafter roof and gable wall.

Roofs and gables built before 1976 are unlikely to have any bracing or straps at all. Quite commonly there will be nothing more than just a piece of skirting board, floorboard or batten that the carpenter used to hold them in place, whilst the roof was erected. Roofs built between around 1976 and 1985 are likely to have some wind bracing and straps. Roofs built after 1985 are likely to be quite securely wind braced.

Provided that the wall and trusses have not moved or distorted much, wind bracing and straps can be introduced to restrain the roof and the gable wall. Over 25 mm out of vertical and a structural engineer should be instructed. Additional bracing in the form of plywood sheets may be used to provide additional rigidity. Over about 100 mm out of vertical and the roof may need to be re-built completely.

If the wall has moved considerably out of vertical, it may need to be re-built. A general rule of thumb is that the wall should not be more than one-third of its total width out of vertical. However, roof tiling normally only over sails the edge of a wall by about 50 mm. The wall and roof junction would probably leak before this degree of movement was reached. In which cases the wall would need to be rebuilt, or the roof altered, at an earlier stage, in order to perform its weatherproofing function.

False Chimneys, Lateral Instability and Movement in Gable Walls

False chimneys made from Glass Reinforced Plastic (GRP commonly known as fibreglass) commonly used from around 1990 onwards, GRP stacks may become detached with time. False corbelled masonry chimneys commonly used between around 1990 and 2010 and also used in extensions to older buildings to maintain architectural appearance. Corbelled stacks commonly lean and create instability in gable walls. Movement is progressive. Remedial work is to take down the corbelled stack and in severe cases rebuild the gable wall.

Key features

1. False corbelled chimney stack brickwork.
2. Chimney leans.
3. Gable wall leans.
4. Horizontal crack (or cracks) in the gable wall.
5. Risk of wind damage to glass reinforced plastic chimneys.

Practical Guide to Diagnosing Structural Movement in Buildings, Second Edition.
Malcolm Holland.
© 2023 John Wiley & Sons Ltd. Published 2023 by John Wiley & Sons Ltd.

Most chimneys in older buildings are not false. Historically chimney breasts were built up from ground level, with the top part of the stack being visible above the line of the roof. They were built there to provide a flue for a fire. Sometimes though, for example when old houses are extended, a false chimney is built to maintain an architectural balance externally.

From around 1990, in the United Kingdom, it became fashionable to build false chimneys on modern houses, to give an older appearance. A mock Victorian or Georgian style was quite common.

False chimneys are usually moulded GRP (Glass Reinforced Plastic, commonly known as fibre glass). GRP chimney mouldings are usually clad with clay brick slips to give a convincing appearance. The false GRP stacks are nailed to the roof timbers. GRP has a high co-efficient of expansion and the stacks are subject to wide variations in temperature, from cold clear night sky to hot summer days. It is easy to imagine that repeated movement, over a period of years or decades, will lead to cracking of the GRP at the fixing points. It is also easy to imagine that the brick slips may lose adhesion from repeated expansion and contraction of the GRP.

The stacks are also in a position where they are exposed to wind. Wind rattling could also cause cracks at fixing points. It is possible (likely?) that GRP chimneys will become insecure with time and could be dislodged and be blown off the roof during storm conditions

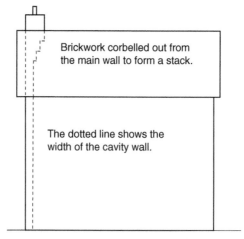

Figure 2.9.1   False chimney corbelled from wall.

When this fashion for false chimneys began, there were probably few suppliers of false chimneys in the market. Instead of using GRP, some developers built false chimneys in brick as an alternative.

To avoid the cost of building chimney breasts at ground level and first-floor level, the brickwork was sometimes corbelled out from the cavity wall, just within the roof space (see Figures 2.9.1 and 2.9.7). Corbelling is stepping out of the brickwork at each course.

In some cases, the corbelled stacks would not be completely false but would serve flue blocks built within the wall. (See Chapter 2.28 Heat Expansion of Flue Blocks.)

The width of a typical cavity wall of this period would be between 250 and 300 mm thick. A minimum stack width would be 450 mm.

The effect of corbelling out the brickwork is to create an eccentric load at the top of the wall. This makes the wall unbalanced at the top.

Over time the eccentric load of the unbalanced stack causes the chimney stack to lean. With time this can bend the top half or entire gable wall (see Figures 2.9.2 and 2.9.4). In modern construction, the roof structure is usually modern lightweight timber roof trusses. The roof structure is wind braced and strapped to the gable wall. The connection of the straps to the gable wall, however, is unlikely to be secure enough to support the wall. The straps, which are hooked into the cavity, will almost certainly slide through the mortar joints of the wall.

As the wall tips over, out of vertical, horizontal cracks may occur at the base of the stack where it corbels out or along the mortar joints of the gable wall (see Figures 2.9.5 and 2.9.6).

The degree to which this happens and the time scale over which it happens, depends on several factors. The most significant is the extent of the corbelling. A 450 mm stack built off a 250 mm thick wall creates a greater eccentricity than a 450 mm stack built off a 300 mm thick wall. The angle of the corbel will also influence it. The greater the step out on each course of brickwork, the more unstable it will be. The length of the wall and whether there are any internal walls at right angles to it, to provide restraint, will be a factor. Orientation, in relation to the prevailing wind over years and decades, is also likely to influence it. If the stack is side on to the prevailing wind, it is likely to tip more and quicker.

Once the stack and gable wall has moved out of vertical, the eccentricity increases and therefore movement can be expected to be progressive. Eventually it might lean on the roof timbers. If the roof is well braced and covered with a heavy roof covering, such as concrete tiles, it might provide enough resistance to arrest it, but this is probably unlikely and

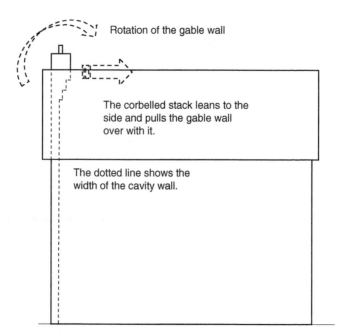

Rotation of the gable wall

The corbelled stack leans to the side and pulls the gable wall over with it.

The dotted line shows the width of the cavity wall.

Figure 2.9.2   Rotation of gable wall.

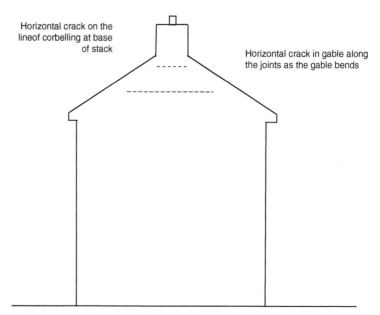

Horizontal crack on the lineof corbelling at base of stack

Horizontal crack in gable along the joints as the gable bends

Figure 2.9.3   Horizontal crack in gable wall.

cannot be relied on. It is possible that the reverse could occur, and the weight of the leaning wall could also push the roof timbers out of vertical.

In modern construction, the first-floor partitions are not usually load bearing. It is therefore unlikely that there would be any way to build up from the internal walls to provide lateral stability and arrest the movement. In this case the only solution would be to remove the corbelled section of the stack.

If this is done soon enough, before the gable wall is pulled out of vertical to any significant degree, it may be enough to resolve the problem. If the gable has also been pulled over, it may need to be partially rebuilt or completely rebuilt depending on the extent of the movement. As the stack is only usually corbelled out in the top half of the gable, the most common scenario would be just the removal of the stack or rebuilding of the top half.

As the manufacture and supply of GRP stacks increased, the use of corbelled false stacks became less common. There is therefore a relatively narrow window of time where this type of construction is quite common. The period is mostly from around 1990 to 2010, although there can of course be exceptions to this. The manifestation of the defect starts from the day it is built but like most things in buildings, it tends to move slowly. Apart from possibly a few exceptions, it would be at least 10–20 years before it became a problem, and probably much longer than that in the majority of cases.

Figure 2.9.4 Leaning corbelled brick stack serving flue blocks built into the inner leaf of the gable wall.

Crack at base of stack where it is corbelled out from the gable wall

Figure 2.9.5   Horizontal crack 9 courses down from the lead tray.

Figure 2.9.6   Close up of crack from Figure 2.9.5.

Figure 2.9.7   Flue blocks and corbelled brickwork at the top of the gable inside the roof space.

2.10

Overloaded Beams

Historically beams were not designed by mathematical means. Beams were sized empirically based on experience. There has always been a desire to build to the practical minimum. Undersized beams deflect under load. A triangular crack often occurs over the opening. Severity is a matter of degree in each case.

Beams over openings, in older buildings, were not designed mathematically, as would be the case today, but were sized by experience and empirical rules of thumb. As a result, they were quite commonly overloaded to some degree. A smaller beam is generally cheaper than a large one and there has always been an economic incentive to design and build to the minimum practical. Designing to a practical minimum size is arguably a good thing. Designing to the smallest practical minimum size is the most efficient solution, as it avoids waste and is therefore more environmentally sustainable.

All materials will deflect to some degree under load, even when they are nowhere near their maximum load bearing capacity. Over time and under load all materials will creep. Timber will bend a lot over time. Steel and concrete will also deflect, but to a lesser extent. So, a beam that is perfectly capable of supporting a load without collapse will sag bit by bit over years, decades or centuries.

Practical Guide to Diagnosing Structural Movement in Buildings, Second Edition. Malcolm Holland.
© 2023 John Wiley & Sons Ltd. Published 2023 by John Wiley & Sons Ltd.

Even in modern buildings where beams have been mathematically designed with generous factors of safety, there is some allowance for acceptable deflection and again long-term creep will occur.

Historically people were willing to accept a much greater degree of deflection than we are today. Look at any old roof and it will visibly sag. Modern roofs are expected to be straight to the naked eye. The result is that timbers, designed in accordance with modern regulations, have to be much bigger than the Georgians or the Victorians would have used; even allowing for the relative strength of the timber available at the time. One way to 'go green' and save cutting down so many trees, would be to relax structural design safety factors. Buildings would not collapse as a result; they would just bend a little more. A little bending or distortion is often considered to be part of the charm of an older property. Many people are willing to actually pay a premium for it in an old building.

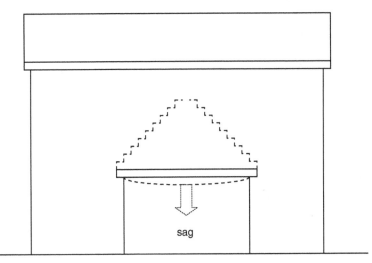

Figure 2.10.1 Typical triangular crack pattern from beam/lintel deflection.

The amount of deflection related to this type of movement will vary depending on the degree of overloading. In most cases cracking will be hairline to a couple of millimetres and no remedial action will be required.

The appearance of the cracks will be stepped (see Figure 2.10.1). The magnitude of displacement will be similar both horizontally and vertically.

Cracking can be much more pronounced than a couple of millimetres before there is any great structural concern. Timber (depending on species) will deflect considerably before it is anywhere near collapse point.

However, when deflection is clearly visible and cracks are 3–5 mm or more, there is a tendency for people to become concerned. It is not just the risk of failure but also how safe people feel within a building that is important. When cracks develop to this degree it would be worth having design check calculations carried out by a structural engineer or take pre-emptive action to strengthen the support.

Similarly, if a timber beam is splitting then strengthen it. If a steel or concrete beam has deflected to this degree, then strengthen it.

On the point of check calculations from an engineer using modern design criteria, it should be recognized that almost all old beams would fail the mathematical calculations. This applies even where they have been perfectly adequate throughout their lives and would continue to be so. This is purely due to the limitations of the assumptions that a designer can reasonably make, and the factors of safety involved in the design. For example, an engineer cannot know how strong an old piece of timber is without testing it to destruction, which would negate the whole point of the exercise. The engineer has to make an assumption. Quite reasonably this assumption has to be on the safe side of the possible range of strength that the timber could have.

This is a limitation of the mathematical approach that cannot be avoided, which is why most old buildings are not assessed in this way. In practice they are assessed visually, using the techniques described in this book or similar, and calling on experience and judgement.

A common example of this type of deflection is found in houses with single-storey bay windows of the late Victorian and interwar period (see Figure 2.10.2). The supporting beam above the bay window is usually built of timber and known as a bressummer beam. With time this deflects. The cracking is often hidden behind the roof of the bay window externally. Internally it is likely to have been decorated over. The deflection can, however, be seen in the first-floor-level window sills above the bay (see Figure 2.10.3). There is also likely to be a crack in the ceiling internally across the bay window opening.

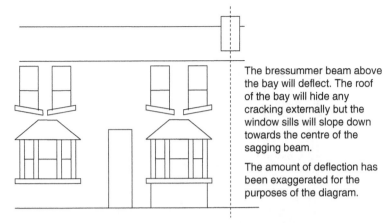

The bressummer beam above the bay will deflect. The roof of the bay will hide any cracking externally but the window sills will slope down towards the centre of the sagging beam.

The amount of deflection has been exaggerated for the purposes of the diagram.

Figure 2.10.2 Deflection of bressummer beam over bay window.

Figure 2.10.3 There is deflection of the timber 'bressummer' beam over the bay window. First-floor window sills have dropped.

Figure 2.10.4 Deflection of a timber lintel (beam) over a window opening.

2.11

Absence of Lintels (Beams) Over Openings in Cavity Walls

The external leaf of the cavity wall is sometimes supported on the window frames, rather than by an independent support. This is particularly the case in houses built between 1945 and 1975. Problems may occur when replacement windows are installed, which are not able to provide support. Whether this defect is significant or not is usually dependent on the size of the openings.

Key Features

1. No visible lintel beneath opening.
2. Horizontal cracking in mortar joints, not bricks just above openings.
3. Quite commonly found in buildings built between 1945 and 1975.
4. Movement usually related to the introduction of replacement windows.
5. Window frames distorted or will not open freely.

Practical Guide to Diagnosing Structural Movement in Buildings, Second Edition. Malcolm Holland.
© 2023 John Wiley & Sons Ltd. Published 2023 by John Wiley & Sons Ltd.

In cavity wall construction the internal leaf of the wall carries the entire load (Figure 2.11.1). The roof and the walls are all supported on the inner leaf. The external leaf of brickwork provides stability to the inner leaf by increasing its effective thickness but the only load that it carries is the self-weight of the bricks.

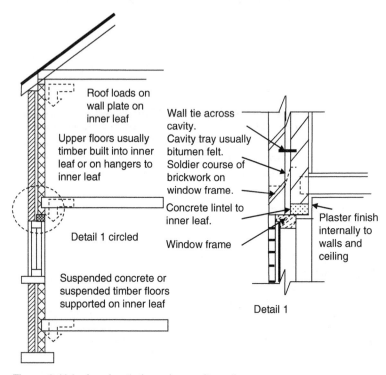

Roof loads on wall plate on inner leaf

Upper floors usually timber built into inner leaf or on hangers to inner leaf

Detail 1 circled

Suspended concrete or suspended timber floors supported on inner leaf

Wall tie across cavity.
Cavity tray usually bitumen felt.
Soldier course of brickwork on window frame.

Concrete lintel to inner leaf.

Window frame

Plaster finish internally to walls and ceiling

Detail 1

Figure 2.11.1 Load path through a cavity wall.

The inner leaf has to be supported over window and door openings by reasonably strong beams, commonly known as lintels. Traditionally these would usually be timber. The use of reinforced concrete lintels began in the interwar period but did not become widespread until after World War Two.

After World War Two, precast concrete lintels became the most commonly used type of support.

From around the mid-1960s, to the mid-1970s, steel lintels became increasingly common, until steel became the virtual universal choice, from around the mid-1970s to date.

The outer leaf only carries its self-weight and does not need much support in most cases. It is not uncommon to find that the outer leaf of brickwork does not have any supporting beam at all, and simply gains support from the window or door frame. Traditionally, timber or steel frames for doors and windows were able to support the loads over most openings, up to around 1.8 m wide. Windows would normally be divided up into separate panes at around 600 mm centres. For example, a 1.8 m wide window would be divided into three panels, with a vertical frame (mullion) at every 600 mm. These vertical supports, being spaced so closely together, were capable of providing support without deflection.

Problems don't usually occur until the original windows are taken out and replaced. When the replacement double glazing industry began, after the oil crisis of the mid 1970s, it was most common to use aluminium for the frames, set in a hardwood surround. In order to keep cost down, it was common to replace multiple bay windows with a window containing just one very large double glazed sheet of glass. Taking out the mullions removes the intermediate support that they used to provide.

Over time plastic windows replaced aluminium as the material of choice for replacement windows. Plastic frames are not normally capable of providing the level of support needed, even if the mullions are spaced in close bays. They can, however, be reinforced internally with steel to provide support.

If there is no existing support to the outer leaf of brickwork, additional support should be provided over the openings when windows are replaced. Alternatively, the replacement frames should be designed to provide support. Unfortunately, in practice it is quite common to find that neither support, nor reinforced frames have been installed.

The absence of support to the outer leaf will not cause structural failure, as it is the inner leaf that carries the building loads. In practice, it is not uncommon for there to be no support to the outer leaf, but even without support no movement occurs. The worst case scenario is that the brickwork above the window opening will loosen and the weight will transfer onto the window frame. If the frame cannot support the weight, it will deflect. Initially this will cause the opening lights to stick. With time the pressure will build up and the glass will crack, and then break completely.

The most common evidence of movement is horizontal cracking along the mortar joints, above the opening. If there is a soldier course of brickwork above the opening (bricks laid standing on their narrow end, like soldiers standing to attention), the cracks are likely to occur at the head of the soldier course. If the brickwork is laid in normal stretcher bond, the separation will usually be in the first and second bed joint, or indeed in both (see Figure 2.11.3). The cracks are quite commonly hairline to 2 mm wide. With time and depending on the width of the opening, they could develop to 5 mm. There will be a few cases where the cracks could develop further, and bricks could possibly even fall out.

Usually there will be a cavity tray built across the cavity within the wall and just above the opening. The purpose of the cavity tray is to discharge any moisture running down the internal face of the external leaf and prevent it causing dampness at the head of the window opening.

The cavity tray is bedded into a mortar joint, and this weakens the mortar joint. A crack on the line of the mortar joint containing the tray is the most likely place for it to occur.

The severity of the cracking that is likely to occur depends on several factors. Firstly, consider the size of the openings. If the openings are quite narrow, say 900–1200 mm, the brickwork will be able to self-corbel over the opening to a large degree. Only a very small triangular section above the opening will weigh on the frame. With such small openings, little or no movement is the most likely outcome.

With larger windows or patio door openings, particularly 1800 mm wide or wider the likelihood or magnitude of movement is increased.

The crack patterns and movement that are likely to occur are demonstrated in Figure 2.11.2.

Where the openings are between 1200 mm and 1800 mm wide, cracking may occur or it may not depending on several factors of design and construction. For example, consider how the load travels through the building and how the brickwork will be able to self-corbel through the angle of the brick bond. Only the triangle of brickwork above the opening will weigh on the frame. The wall ties in a cavity wall will also provide some resistance and help to hold up the outer leaf, by transferring load to the inner leaf. Metal wall ties are not designed to do this, but they will to some degree.

The alignment of openings and the load path is significant. If the first-floor openings are not aligned with the ground-floor openings, a load concentration could occur, or natural corbelling may be prevented to some degree.

The quality of the brickwork is also a factor. The brickwork is wedged in place by the nature of its construction. If the brickwork is good quality, with perp joints (vertical joints) completely full of mortar, it will be wedged in quite tightly. If, however, the brickwork has been not so well laid, and the perp joints are not completely full, the wedging in effect will be less. In Figure 2.11.4 the window opening is approximately 1500mm wide. It has deflected a little but no significant cracking has occurred.

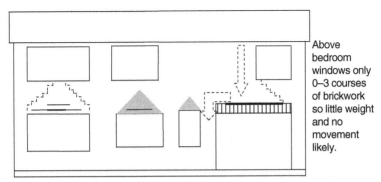

Above bedroom windows only 0–3 courses of brickwork so little weight and no movement likely.

Over large openings, corbelling action breaks down and cracking is likely to be more pronounced, with horizontal cracking on bed joints and stepped cracking at 45 degrees developing.

Area of brickwork on window marked grey. Natural corbelling through the bond of the brickwork can bridge over small openings without cracking. The wider the opening the more likely it is that brickwork will drop and crack on the bed joint

Crack pattern likely to be more pronounced and greater in size if load path exacerbates the deflection.

Figure 2.11.2 Absence of lintel support over openings in cavity walls.

Unfortunately, most brickwork is laid at a price per thousand. There is an incentive to lay fast rather than well. It is down to the management of quality control on site to ensure that speed does not compromise quality. Local authority housing of the 50s and 60s is usually quite good in this respect, because a clerk of the works was commonly used to inspect the works as they proceeded.

The quality of private houses, built for profit throughout the 60s and 70s, is often quite suspect in this respect. Properties built during the house price boom of the mid to late 80s may also be quite variable, as there was a great deal of pressure to get them built and sold in a short period.

Movement is more likely to occur on the south and west elevations. Seasonal movement is greater on these elevations due to exposure to the sun and driving rain. A leak through any cracks over the openings is likely to be greater on the exposed west or southwest elevations.

There will also be a tendency for the movement to 'creep' over time but this could be measured in years or decades, depending on the factors already discussed.

This defect most commonly occurs in houses built after World War Two, up to the mid-1970s. Commonly, there will be a soldier course of brickwork above the windows. Another indicator is a visible cavity tray in the outer leaf, at the head of the soldier course, or two-three courses above the opening if the brickwork is stretcher bond.

Remedial work is a matter of judgement based on the degree of movement and the likelihood of it getting worse with time. In many cases, possibly even most cases, no work will be required in the short to medium term, other than perhaps some minor re-pointing or sealing with mastic. This is to prevent the risk of rainwater penetration on exposed elevations.

It is difficult to put in new lintel support to the outer leaf, without taking out the windows or risking damage to them. It will always be impossible to stitch in new brickwork to match the existing. If all the windows are working and there is no water penetration, the works can usually wait, possibly for years and decades. The hope would be that works could be delayed until it next becomes necessary to replace the windows, which could be in 30 years' time.

Figure 2.11.3 It is just possible to see that the brickwork has dropped slightly at the left side and there is just a hairline horizontal crack along the bed joint two courses above the opening.

Although this approach might be acceptable from a strictly technical point of view, it is sometimes not acceptable to transfer a risk into the future. This is particularly true when a building is changing ownership. The potential future owner may not be willing to take over the risk from the current owner, without some monetary compensation. Such arguments are, however, beyond the intended scope of this text.

Figure 2.11.4   No lintel above the opening. The soldier course is supported on the frame and there is a tile creasing and cavity tray above that. There is very slight deflection but no significant movement or cracking. The window opening is 1500 mm wide.

2.12

Overloaded Floors

Historically floor joists were not designed by mathematical means. Joists were sized empirically based on experience. Joists are often small by modern standards. Some deflection or springiness is common. Local overloading under the weight of partitions or load concentrations is also common. When a change of use occurs, floors may become more heavily loaded than they were originally designed for. Whether this defect is serious or not, is a matter of degree on the facts in each case.

Key features

1. Deflected, sagging floors. May be localized if uneven load.
2. Springy floors.
3. Door openings out of square, doors may bind or have been cut out of square to fit.
4. Separation of internal walls from external walls, widest at top.
5. Separation from the ceiling. Cracks between the wall and the ceiling will be widest at mid span.
6. Ceilings crack beneath.
7. Crack sometimes between door head and ceiling.

Practical Guide to Diagnosing Structural Movement in Buildings, Second Edition. Malcolm Holland.
© 2023 John Wiley & Sons Ltd. Published 2023 by John Wiley & Sons Ltd.

General overloading of floors is often caused when old residential buildings have undergone conversion and change of use to office use. Office use commonly involves much heavier floor loading in the form of furniture and storage.

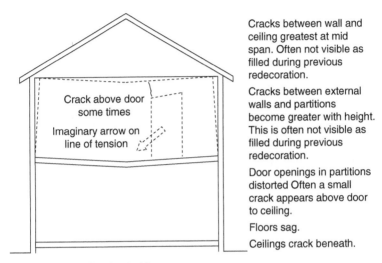

Cracks between wall and ceiling greatest at mid span. Often not visible as filled during previous redecoration.

Cracks between external walls and partitions become greater with height. This is often not visible as filled during previous redecoration.

Door openings in partitions distorted Often a small crack appears above door to ceiling.

Floors sag.

Ceilings crack beneath.

Figure 2.12.1 Overloaded floors.

More commonly local overloading of floor joists is often caused when partition walls are built off upper floors. The joists under a partition should be doubled up. Quite often this is omitted.

Historically, prior to the last 20–30 years, it was more common to find that joists had not been doubled up under non-load bearing partitions. As a result, deflection can be pronounced, often 15–25 mm.

When this type of movement is suspected, it is important to record which way the floor joists span. If a partition goes across the span, the weight is shared by several joists and deflection will be small. If the weight of partition is built on one joist, in the direction of the span, the movement will be greater. The floor level can be checked and measured using a spirit level.

If there are door openings within a wall, which is built off a deflected floor, the door opening will be out of square. It is very simple to use the distortion of a door opening to show where and how the movement has occurred, in the same way as the 'square' distortion described in 'First Principles' at the beginning of this book. It is reasonable to assume that the door opening was built square. The distorted new shape can be sketched to see which side has dropped. It is then possible to see which diagonal has been stretched. Follow the imaginary lines of tension from this diagonal and they will point to where the movement has occurred (see Figure 2.12.1).

This type of movement is very common. It occurs in nearly all buildings, where the first-floor walls are built off the floor. In pre–World War Two houses, it is very common to find that the floor plan is, for the most part, repeated at ground-floor level and at first-floor level. The masonry first-floor walls are often built off the ground-floor walls. Where there are timber frame partition walls at first-floor level, they are likely to be built off the floor.

After World War Two, designers moved away from aligning first floor walls with ground-floor walls. It is quite common, in houses built between 1945 to the late 1960s, to have even masonry walls built off the floors at first-floor level. Deflection in such circumstances can be pronounced. From the 1950s onwards, masonry first-floor walls declined in usage. Lightweight first-floor partitions became increasingly popular, and almost universal from the 1970s onwards. These partition walls are plasterboard lined for fire resistance, but the core of the partition could be a lattice of cardboard, compressed strawboard or timber framing.

In addition to partition walls being built off the floors, there are also other loads that commonly cause deflection. The most common being the weight of the water cylinder in the airing cupboard. The weight of water in the bath may also cause deflection.

Load concentrations are also a very common cause of deflection, for example, where the roof is strutted to an internal wall that is built off a floor.

Where openings have been formed in the floor, for example to form the staircase opening, it is also common to find some deflection, if trimming timbers have not been doubled up.

Timber is an elastic material which will inevitably stretch and deflect under load, over years and decades. In virtually all cases the deflection of the floor will be less than 25 mm. Whilst this will be noticeable, it will not usually be structurally significant, provided it has occurred over decades or centuries. In fact, sometimes it is considered to be part of the charm of an old building. In such circumstances no remedial work is required.

If the movement is greater or has occurred quickly then remedial action could include:

1. Strengthening the floor with additional joists.
2. Reducing the span of the floor by introducing a beam at right angles to the span.

3. Reducing the load, for example by replacing masonry walls with light partitions.
4. Reducing loads by reducing furniture, particularly storage cabinets or similar.
5. Locating heavy loads near to support points.

2.13

Overloaded Walls

It is relatively rare to find internal walls that are overloaded. There are, however, instances where internal load bearing timber frame walls are compressed under long-term load. When a change of use occurs floors may become more heavily loaded than they were originally designed for. If floors are supported on walls that are overloaded, the deflection can cause the floors to slope. Whether this defect is serious or not is a matter of degree on the facts in each case.

Key features

1. Sloping floors. Slope increases at each floor level.
2. Separation of internal and external walls, separation increases at each floor level.
3. Distortion of door openings.
4. If the ceiling structure can span to the external walls or is well supported on hangers to the purlins or rafters, then the walls may separate from the ceiling at top floor level.

Overloading of internal walls is relatively rare. For the most part, masonry buildings are of a scale where the walls are strong enough and thick enough to support the loads imposed on them. There are, however, some exceptions as the number of floors increases to above two storeys.

Practical Guide to Diagnosing Structural Movement in Buildings, Second Edition. Malcolm Holland.
© 2023 John Wiley & Sons Ltd. Published 2023 by John Wiley & Sons Ltd.

Internal load bearing walls in a two-storey dwelling used to be usually half a brick thick (112 mm). After World War Two, 100 mm thick concrete blocks began to replace bricks and soon completely replaced the use of bricks for internal walls. Where a masonry load bearing wall is three storeys in height, the wall at ground level should be at least 190 mm thick. Before concrete blocks became the norm for internal walls that would normally be exceeded, by building the wall a full brick width (215 mm). Now it would be 200 mm thick blockwork.

Historically, however, this rule was not always followed. It is relatively common to find three-storey high internal walls that are only half a brick thick for their full height. Such thin walls can be compressed slowly with time, particularly at load concentration points. This compression can lead to bowing of the wall.

Internal walls may also be timber frame. This can be the historic oak frames with infill panels, originally wattle and daub in historic buildings. In Georgian and Victorian buildings, it is commonly 100 mm × 50 mm timber studs fully enclosed and finished with lath and plaster. In the 1930s, plasterboard began to replace lath and plaster. After World War Two, plasterboard became the normal lining material for timber studwork partitions. In the 1990s, the common use of electric screw drivers saw the introduction of metal studs as an alternative to timber studwork, with screw fixed plasterboard, rather than nailed boards.

Whilst the vast majority of internal timber frame walls are not load bearing, it is not that uncommon to find load bearing timber frame walls. In modern timber frame construction, all the load bearing walls are of course, timber frame, with studs usually nominally 100 mm × 50 mm at 600 mm centres.

Timber is an elastic material and under load it will be compressed with time. If, for example, the design loads are increased by change of use from domestic to office, the risk of this increases. There are, for example, numerous cases where Georgian and Victorian three or four-storey town houses have been converted to office use as the commercial centres of towns have developed to swallow up what were historically the wealthy professional suburbs of the period.

Many of the symptoms are very similar to deflection of floors. With deflection of floors, the amount of separation will not increase with each successive floor. Where it is related to compression of walls stacked on top of each other, the movement will become greater with height. With floor deflection there will be separation from the ceilings on each floor and this will be greatest at mid span (see Figure 2.13.1). With compression of the walls there will not be separation from the ceiling, apart from perhaps the ceiling at loft level, if this is supported on trusses or securely hung from the roof structure.

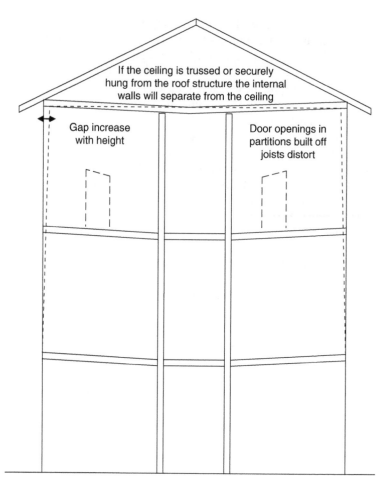

Figure 2.13.1 Overloaded walls.

Some aspects of this type of movement can also easily be confused with roof spread. In both cases there will be separation between the internal and the external walls. In both cases this will increase in magnitude with height.

It is possible to differentiate between the two by noting several factors. With roof spread the external walls will be pushed out of vertical to lean outwards. There is no reason for this to occur where the internal walls have been compressed. In some cases, the external walls could even have been dragged inwards by internal walls, built into them at right angles. Measure the verticality of the walls with a spirit level.

If the walls are wall papered, check around the corners between the internal and external walls. Record whether it is ripped or rippled.

Roof spread stretches the paper outwards and will cause it to rip. Where the floors have gone down and taken the partition walls at right angles to the external walls down with them, this will have dragged the wall paper causing it to ripple. The angle of the ripple is the angle of compression and opposite to the angle of tension. If the angle of compression is determined, the tension is at right angles to it. Once the direction of the tension has been determined it is possible to imagine the 'arrows' of tension which will point to the direction of the movement.

2.14

Differential Movement

> *Differential movement occurs at the junction between different materials, constructions or ages within a building. It is not structural and is minor in nature.*

Differential movement can be foundation related or non-foundation related. Differential movement related to foundations is discussed later in the book.

Differential movement not related to foundations may be caused by several different things, but it is the junction of two different things in each case. For example:

1. A junction between different ages of a building, for example, a modern extension to a Victorian building. (See Figure 2.14.1.)
2. A junction between two different materials, for example, blockwork and brickwork. (See Figure 2.14.2.)

Typically, cracking will be hairline, but it could be 1–2 mm in some cases depending on the factors involved. Such movement tends to be seasonal or diurnal. If the movement is re-pointed, it will recur. Whilst a crack may recur it will not be progressive in any one direction and will not get worse with time. Usually, therefore, no action is essential.

Practical Guide to Diagnosing Structural Movement in Buildings, Second Edition. Malcolm Holland.
© 2023 John Wiley & Sons Ltd. Published 2023 by John Wiley & Sons Ltd.

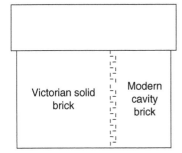

Figure 2.14.1 Differential movement related to different ages.

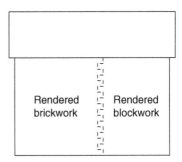

Figure 2.14.2 Differential movement related to different materials.

The only way to stop the crack recurring would be to build in a flexible joint, and this is likely to be more obtrusive than the crack itself, unless it could be hidden, for example behind a down pipe or even a false downpipe. Rainwater downpipes are common on buildings and people are used to seeing them. People tend to 'look past' downpipes and they do not look out of place. In effect they become to some extent, invisible. Whereas people are not used to seeing mastic filled expansion joints on the elevations of buildings. Peoples' eyes are drawn to them as an anomaly, and they look out of place and ugly. It is a common device of designers to put a down pipe in front of a joint to hide it. In some cases, the down pipe might be cut to fit around the gutter as if it was connected, but not actually be connected to it for the purpose of drainage. In this context that is what is meant by the term 'false downpipe'.

Arch Thrust and Arch Flattening

> *Arches can be used to provide support over openings. True half round arches transmit the loads around the openings in compression. When an arch is flatter than a full half circle, a thrust is created at the supports. The flatter the arch is, the greater the thrust will be. The thrust has to be resisted by the mass of the masonry either side of the opening. If the arch deflects, a triangular crack often occurs above the opening. Whether this defect is serious or not is dependent on the facts in each case.*

Key features

1. Brick or stone arches over openings.
2. Loads transferred in compression.
3. Low arches create a thrust force at supports.
4. Thrust resisted by mass of masonry either side of opening.
5. If resistance is inadequate the arch flattens and 45-degree cracking occurs above.
6. Severity depends on degree of movement in each case.

One way of providing support over an opening in a wall is to use an arch. A true full half circle arch will transfer the load around an opening and will not create thrust. The load will be turned through the arch to

Practical Guide to Diagnosing Structural Movement in Buildings, Second Edition. Malcolm Holland.
© 2023 John Wiley & Sons Ltd. Published 2023 by John Wiley & Sons Ltd.

complete its journey vertically. Look at old Roman arches or more modern examples of Victorian railway arches.

These completely 180° arches, use the compressive strength of masonry material and avoid tension forces (Figure 2.15.1).

However, in so many cases, a true 180° arch is not used. Far more commonly, a flatter arch is employed. A flattened arch is not capable of turning the load through the diameter of the circle to act vertically again. As a result, the force of the load tries to push outwards.

The degree to which this happens depends on how flat is the arch. The flatter the arch, the more outwards thrust will be created. If there is an insufficient mass of masonry either side of a flattish arch, the thrust will be able to push the opening wider and the arch will drop (see Figure 2.15.3). If the arch drops and flattens, this will allow the masonry above it to drop, in much the same way as it would with lintel deflection (Figures 2.15.2 and 2.15.4).

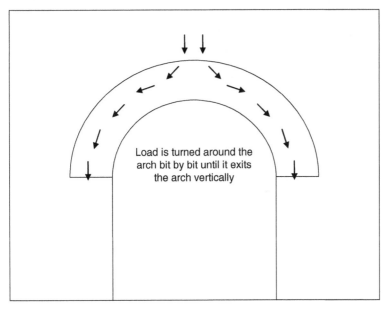

Figure 2.15.1 Load path around an arch.

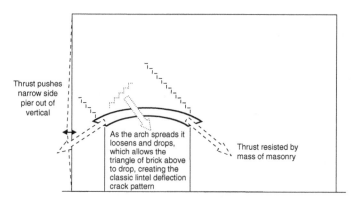

Thrust pushes narrow side pier out of vertical

As the arch spreads it loosens and drops, which allows the triangle of brick above to drop, creating the classic lintel deflection crack pattern

Thrust resisted by mass of masonry

Figure 2.15.2   Arch thrust diagram.

In order to prevent the arch from flattening out and collapsing, there has to be a sufficient mass of brickwork either side of the opening, to resist the outwards force. As a rule of thumb, the width of the pier of brickwork, at the side of the opening should be equal to at least half the width of the opening.

The amount of movement that will occur will be a matter of degree depending on the facts in each case. The amount of movement will depend on how flat is the arch, how wide is the span and how large is the pier. The movement also tends to be very slow and will occur by progressively creeping over decades or centuries.

In most cases these arches were built on empirical rules and the experience of the bricklayer/mason. Collapse is rare, but movement is likely to be progressive, although at a very slow rate.

There are a number of ways to arrest the movement:

1. Increase the size of the pier by buttressing.
2. Replace the arch with a lintel to transfer the loads vertically at the supports.
3. Build in a steel tie rod behind the arch from flank wall to flank wall.

Arches may also flatten to some degree, even if the piers are able to resist the thrust. Even solid, rigid materials like brick and stone will creep and compress to some degree, over decades or centuries. Again the amount of flattening will depend on how shallow is the arch. The flatter the arch, the greater the thrust pressure at the supports and the greater the degree of creep and flattening. Where the buttressing is adequate and the only movement is a little creep movement, this is likely to be only hairline or 1–2 mm in magnitude. Whilst such minor movement might be progressive, the magnitude is so small that it is not significant for practical purposes in most cases.

Figure 2.15.3 Arch thrust photograph. A typically very minor crack over top right hand corner of the window in a 1900 rear addition. The window is close to the corner. There is insufficient mass of brickwork at the corner to resist the thrust.

Figure 2.15.4   Dropped arch. A very wide flat arch in relation to the width of the brickwork at each side. It has pushed out and dropped.

2.16

Arch Spread in Chimneys Built over Alleys in Terrace Housing

This defect is mostly found in 18th century, 19th century and early 20th[th] Century terraced housing, mostly in urban areas. Alleys run through the terrace, with chimneys built centrally over the alleys. The substantial weight causes arch spread. This pushes out the alley walls. Stepped diagonal cracking occurs over the arch. Typically, 1–3 mm. The weight also causes settlement (see Chapter 3.6 Uneven Loading). Typically, movement takes place over the first few decades of the buildings life. No remedial action is usually needed.

Key features

1. Terrace housing with access alleys through the terrace.
2. Chimneys built on brick arches over the alleys.
3. Arch spread and stepped diagonal cracking over the arch. Typically, 1–3 mm.
4. Displacement in the plane of the wall in line with the arch and chimney breast.
5. Vertical displacement from weight, differential loading.

Practical Guide to Diagnosing Structural Movement in Buildings, Second Edition. Malcolm Holland.
© 2023 John Wiley & Sons Ltd. Published 2023 by John Wiley & Sons Ltd.

In the United Kingdom it is common to find houses built as long terraces; particularly in towns, where large numbers of houses were needed to accommodate workers for industry and commerce. This is particularly common in the later part of the 18th Century, throughout the 19th century and early part of the 20th century prior to World War One (1914). During this period, houses were heated by open solid fuel fires (coal in urban areas).

When long terraces of housing are built, one of the technical problems to overcome, is how to gain access to the rear, without going through the house. Sometimes, access was made by rear service roads or rear alleys. An alternative was to run alleys through the terraces at ground floor level, beneath the first-floor bedrooms

When an alley is run through a terrace, the designer has several options about how to divide the properties at the party wall position. Sometimes the first-floor party wall and the chimneys are positioned centrally above the alleys. (See Figure 2.16.1.)

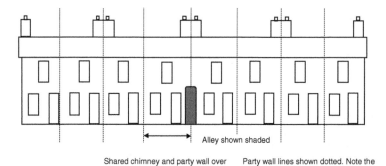

Alley shown shaded

Shared chimney and party wall over   Party wall lines shown dotted. Note the
the alley   position of the shared chimney stacks.

Figure 2.16.1   Front elevation showing the position of the alley running front to rear through the terrace.

Positioning the chimneys and party walls in this way, keeps the terrace symmetrical.

Positioning of the stacks over the alley also allows the chimney stacks to be built as semi-detached and shared. This is cheaper than building an individual stack for every house. Please note that only the stacks are shared, not the flues within the stacks.

It is relatively easy to build the party walls over the alley. The party walls can easily be supported on joists spanning over the alley. However, building the chimney breasts over the alley presents a much greater structural challenge due to their size and weight. (See Figure 2.16.2.)

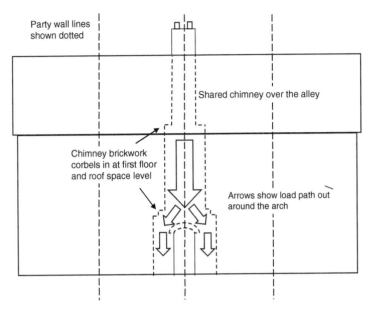

Figure 2.16.2 Chimney built over the alley, and load path around the arch.

The chimney breasts at ground level have to corbel over to the centre line of the alley. Brick arches are constructed over the alley to carry the large weight of the brick chimney stacks at first floor level, up to and beyond the roof level.

As described in the previous Chapter, 2.15 'Arch Thrust and Arch Flattening', the substantial weight on the arch causes thrust.

The thrust pushes the wall out on either side of the arch and causes the arch to drop. As the arch drops, the load path from the brickwork above travels through the corbel of the brickwork. The result of this is that a stepped crack occurs over the arch, just as it would if the arch dropped over a door or window opening. The cracking over the arch is also likely to be quite small, probably around 1–3 mm. The horizontal component of the crack is likely to be greater than the vertical component because the movement is caused by outwards thrust. The comparative movement will however vary depending on the flatness of the arch and how much settlement has also occurred at the support points. The flatter the arch, the more the movement will be horizontal.

If there is enough outwards thrust, the wall of the chimney breast in the alley, will be pushed out of vertical either side of the arch.

There will be some slight displacement on the plane of the wall, where the chimney breasts have been pushed a little out of vertical by arch spread. This is likely to be minor, probably no more than a few millimetres or up to about 5 mm. The displacement will be greatest at the top, and taper, to zero at the bottom. It can be measured with a spirit level.

Sometimes the arches will be constructed as full 180-degree half circle arches, sometimes called 'Roman' arches. The reason for creating a full 180-degree arch, is to transfer the load around the arch and back to a vertical direction. This turns the load around the arch and prevents thrust from pushing out the walls. This is the intention, but it is not always effective.

The weight of the chimney brickwork is substantial and is concentrated on the chimney breast positions in the alley walls. The weight causes the chimney breasts to settle downwards more than the other parts of the walls. The two sides may not settle exactly uniformly. If the arch tips a bit to one side, it may allow a little arch spread, as the thrust will be different on each side.

As the chimney settles, a vertical shear crack occurs between the chimney breast and the party wall. See Chapter 3.10 Differential Chimney Settlement.

The wall would have cracked internally as well, but the internal crack would most probably have been filled and decorated numerous times over the lifetime of the building, and there is unlikely to be any evidence of it.

As the chimney breast section is much heavier than the rest of the alley walls, there may be some visible vertical displacement of wall at the junction of the crack. The vertical cracking is likely to be minor, hairline or 1–2 mm. Check if the horizontal mortar joints are out of vertical alignment across the cracks. Sometimes however, the vertical displacement is too small to be clearly seen given the normal irregularities of the brick laying.

If there is a vertical crack and if the chimney breast part of the alley wall has been pushed out of vertical, there can be a step in the plane of the wall at the crack junction (see Figure 2.16.3 and Figure 2.16.4). Again, this is likely to be quite minor, 1–3 mm.

This form of movement takes place over the first few decades. After a few decades, the movement stops, for all practical considerations. There may be some long-term creep, but this would be of such little magnitude and over such a long period of time; that it would be insignificant. It can therefore be considered as historic in nature. No remedial action is required.

If the crack is re-pointed on the alley side, it is likely to recur as a hairline crack. This is simply due to differential movement, daily or seasonally.

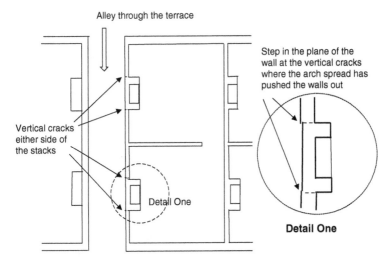

Figure 2.16.3 Walls pushed out of vertical on the chimney breast line by arch spread.

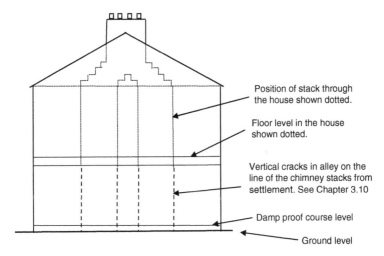

Figure 2.16.4 Vertical cracks on chimney breast line.

Buckling of Jambs to Sash Windows and Arch Flattening

> *Sash window openings in 18th-century, 19th-century and 20th-century buildings often have staggered masonry jambs (sides) creating narrow half brick nib piers. Arches are combined with timber lintels for support over the openings. The half brick staggered jambs are vulnerable to buckling under load and arch thrust. Buckling eventually changes the distribution of the load path and movement stops. Buckling is a common cause of overall wall shortening and other associated movement.*

Key features

1. Openings for sash windows with rebated brick jambs and brick arches above.
2. Buckling and bowing of brick jambs.
3. Possibly stepped cracking over the arch.

In most typical arch constructions over windows and doors, in Georgian and Victorian buildings, the opening is only partly supported by an arch. There is also a timber lintel behind the visible arch.

Typically, a sash window will be positioned behind the brick arch, beneath a timber lintel at the head of the opening (see Figure 2.17.1). The window frame is set behind a brick nib at the jambs (sides).

Practical Guide to Diagnosing Structural Movement in Buildings, Second Edition.
Malcolm Holland.
© 2023 John Wiley & Sons Ltd. Published 2023 by John Wiley & Sons Ltd.

With normal sash window construction, the frame of the sash window is positioned behind the jambs of a staggered brick jamb. This leaves a narrow half brick nib on which the arch at the head is supported. Behind the arch at the head, there is a flat timber lintel hidden by the internal wall plaster and timber lining and architrave. The window frame is set behind the brick arch and up to the timber lintel. The arch and the timber lintel share the load.

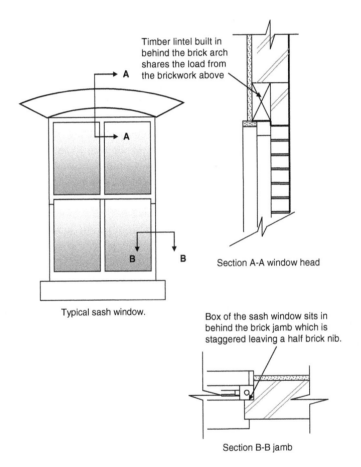

Timber lintel built in behind the brick arch shares the load from the brickwork above

Section A-A window head

Typical sash window.

Box of the sash window sits in behind the brick jamb which is staggered leaving a half brick nib.

Section B-B jamb

Figure 2.17.1   Sash window-opening construction.

Timber is a more elastic material than brickwork and it will bend under load with time. As this occurs, the load will be disproportionately transferred onto the brick arch. The brick arch transfers the load onto the narrow nib of brickwork. The half brick nib is, in effect, a narrow pier and under long-term compression this slender pier of brick will bow. It is very common to observe a gap between the sash window frame and the bowing brickwork at the jamb. As the narrow pier bows, the arch flattens and drops. As the arch drops, it transfers the weight back to the timber lintel. The timber lintel then bends, and the process repeats itself. With time the nib bows to such a degree that the load path by-passes the half brick nib and the weight is able to spread out into the wall, where it is a full brick thick. The full brick thick part of the wall is strong enough not to buckle and is able to resist any further thrust. So once the half brick thick nib has been buckled out of the way, the movement stops.

There will be bowing of the narrow pier at the jambs of the opening. This bowing will usually be quite small, say 2–5 mm, but it often looks much worse because the mortar may fall out of the joint. If typically, a mortar joint is about 10 mm wide, the result could be a gap of 15 mm between the window frame and the brickwork (see Figure 2.17.2).

Buckling of the jambs causes the height of the wall to shorten either side of the window. This can set in process a quite complicated series of events which is described later in Chapter 2.18 Wall Shortening and Sequential Movement in Walls and Roofs.

As the jambs buckle and bow, the brick arch over the opening will flatten and spread. The degree of movement is usually quite small and takes place slowly over decades. As the mortar in a house of this age is usually a relatively soft lime mortar, it is common to find it can accommodate this small gradual movement without cracking. It creeps slowly and is flexible enough to absorb the movement.

In some cases, cracking will occur, and the crack pattern will be very much the same as with lintel deflection, a 45° triangle of cracking from both sides of the opening (see Figure 2.17.3).

Figure 2.17.2 Buckling at sash jamb. The single brick thick nib at the jamb of the sash window has buckled a little and the mortar has fallen out. The movement looks much worse than it actually is because the mortar has fallen out.

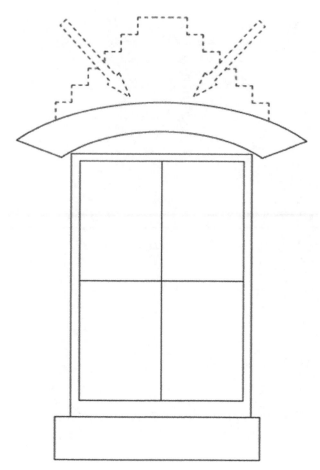

Figure 2.17.3 Crack pattern above a dropped arch.

Wall Shortening and Sequential Movement in Wall and Roofs (Holland's Multiple Factor Diagnosis)

Sometimes found in terraced housing. This is a complicated sequence of events dependent on multiple factors. The jambs of sash windows bow. This buckles the wall. Possible deflection of lintels or bressummer beams. This causes deflection. The height of the elevation is reduced by buckling and deflection This causes reduction of roof support at the eaves. The roof spreads and bows the wall. The elevation is pulled down and to the centre, away from the party walls. Separation cracking occurs between the front elevation and the party walls. Staggered cracks occur in the party walls, near the front, at upper level and extend into the roof space. Evidence of internal cracking is often not visible because of internal redecoration over the life time of the building. If present the restraining effect of rear additions prevents movement in the rear, so movement will be just to the front elevation in most cases. Movement is non-progressive after the first few decades. Cosmetic repairs only.

Practical Guide to Diagnosing Structural Movement in Buildings, Second Edition. Malcolm Holland.
© 2023 John Wiley & Sons Ltd. Published 2023 by John Wiley & Sons Ltd.

Key features

1. Buckling at brick arches to sash windows buckles the elevation overall.
2. Deflection of bressummer beams over bay windows (if present).
3. Reduction in wall height in relation to buckling (and possibly deflection as described above).
4. Sagging of roof and associated roof spread.
5. Walls pushed out and bow by roof spread.
6. Measure buckling with spirit level.
7. Rear additions (if present) prevent movement so usually front elevation only.
8. Wall elevation pulls inwards away from party walls.
9. Cracking between front elevation and party walls, easily confused with roof spread.
10. Cracking in top part of party wall, extending into roof space.
11. Expect evidence of cracking to be covered by repairs during redecoration cycle.
12. Cracking usually only visible if not decorated for years or inside cupboards.
13. Usually not serious and non-progressive after first few decades.
14. Cosmetic repairs only.

The progression of this movement is quite complicated and involves the interaction of several elements together. Before reading this chapter, please revise previous chapters on:

Chapter 1.1. First Principles

Chapter 2.5. Roof Spread

Chapter 2.10. Overloaded Beams

Chapter 2.17. Buckling of Sash Jambs and Arch Flattening

This sequence of this movement may vary to some degree from case to case. There are multiple factors that can influence the way the movement progresses. Not all factors will be present in every case. The magnitude of each factor may also vary.

The movement is most commonly found in 18th-, 19th- and early 20th- century terrace houses, built with solid walls and with sash windows (Georgian, Victorian and Edwardian periods). There is, however, the possibility that similar factors will occur in houses of other ages or other styles, semi-detached or detached.

Although this form of movement is common, it is not usually recognized, and the diagnosis does not appear in any other textbook on the subject of structural movement. This is because the movement takes place over the first few decades of the building's life and then stops. Internally all evidence of it is usually concealed by numerous redecoration cycles over the life of the building since the movement stopped. The magnitude of the movement would not normally be significant, and no remedial action is normally required.

There are several reasons why the walls of a house may shorten. In the context of this movement, the most common cause is buckling of the sash window jambs at first-floor level (and ground-floor level if present). At first-floor level there is likely to be one or two sash windows. The jambs of the sash windows will have bowed to some degree. The reason for bowing of the sash window jambs is explained in Chapter 2.17 Buckling of Sash Jambs and Arch Flattening.

Some houses of this age may have a bay window at ground-floor level. Over the bay window there will usually be a built-up timber beam (lintel), known as a bressummer beam. Over decades the timber beam will slowly sag under load. This is explained in Chapter 2.10 Overloaded Beams.

Figure 2.18.1 shows the front elevation of a typical row of a terrace of houses.

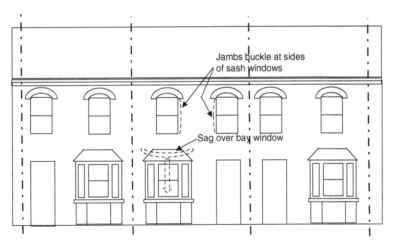

Figure 2.18.1 Typical front elevation of a terrace of houses.

Please note that in the diagram above, the jambs are shown bowing sideways. In reality the brickwork bows outwards (out from the page in the diagram). Unfortunately, however this can't be shown in this two-dimensional diagram. Please refer to Chapter 2.17 Buckling of Sash Jambs and Arch Flattening, where other diagrams and photographs show how the brickwork bows.

The buckling and bowing of the brickwork at the jambs of the sash windows (and sagging of the bressummer beam over the bay if present), will shorten the height of the wall, in line with the openings. If there are two sash windows, the brickwork between the windows will bow more than if there is just one window, as it will be dragged on both sides. The bowing movement occurs over the first few decades of the buildings life and then stops. It ceases to be progressive.

The effect of this bowing of the jambs and lintel sagging is shown in Figure 2.18.2. The centre section of the wall becomes shorter. **Height 'B'** becomes shorter than the original **Height 'A'** shown on the diagram.

Figure 2.18.2   Shortening of the elevation.

As the wall shortens, it takes away support from the roof at the bottom of the rafters, at eaves level. This causes the roof to sag and push out on the walls at the top. In order to appreciate this, one has to know the basics of how a traditional roof is built (see Figure 2.18.3).

Timber roofs are made up with evenly spaced rafters which are supported on the wall at the bottom of the slope. The rafters slope up to the top of the roof where they meet the ridge board.

The ridge board does not support the rafter. The rafter leans against the ridge board and is nailed into it. The rafter on the opposite roof slope leans against the ridge board from the other side and balances it. The ridge board is just there to lean the rafters against for stability. It is a flimsy plank of timber only 25 mm thick, commonly.

Between the top and the bottom of the roof slope there is a horizontal timber beam (purlin) which provides support and prevents the rafters from sagging. The purlins are usually quite large pieces of timber which span between the party walls.

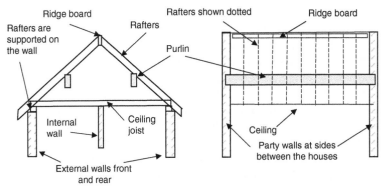

Section front to back through a roof   **Section side to side through a roof**

Figure 2.18.3   Traditional rafter and purlin roof construction.

Across the bottom of the rafters there are ceiling joists. The ceiling joists are not just there to support the ceiling. They are (or should be) nailed to the rafter feet to form a triangle of roof timbers which prevent the rafters from spreading out' (see Chapter 2.5 Roof Spread).

When the wall shortens, as previously described, it takes support away from the rafters at the bottom. (See Figure 2.18.4.)

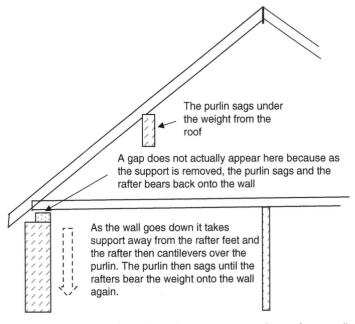

Figure 2.18.4   Walls shortening takes support away from rafter at wall.

The rafter then starts to cantilever over the purlin. This transfers weight away from the wall plate, onto the purlin. This shift in weight, onto the purlin, causes it to sag.

When it sags sufficiently the rafter sits back on the wall by gravity and transfers the weight back evenly between the rafter and the purlin.

The diagram above shows a gap between the wall and the rafter/ceiling timbers. This is for illustration purposes to explain the process. In reality an actual gap does not occur. The movement occurs as a smooth linear transition. It all moves together to the naked eye. The movement takes place slowly over the first few decades of the building's life.

As the purlin sags, it allows the roof to spread a little (see Chapter 2.5 Roof Spread).

The roof spread pushes the top of the wall outwards.

The bowing of the jambs tips the walls inward at window head height (see Figure 2.18.5).

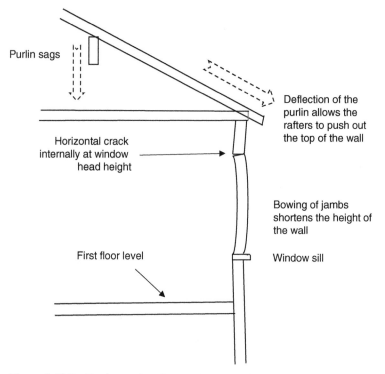

Figure 2.18.5   Bowing and roof spread in the wall.

If there are two-bedroom windows in the front elevation, a horizontal crack is likely to occur at window head height between the two windows. This crack may also extend to the sides of the windows where it will taper out towards the party walls. If there is just one window in the front elevation, a crack is likely to occur either side of the window where it will taper out towards the party walls.

Please be aware that evidence of the horizontal cracks internally may not be visible because it will probably have been covered by repeated redecoration every few years.

The bowing of the walls and the roof spread will be so small that it will probably not be visible to the naked eye. The verticality of the wall can however be measured using a spirit level. A long spirit level will reveal that the wall is not straight, A short spirit level can be used to plot the variations either side of vertical (see Figure 2.18.6).

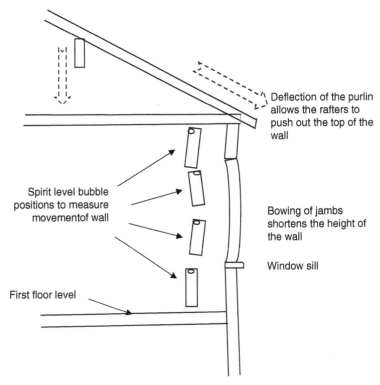

Figure 2.18.6 Spirit-level bubble positions to plot movement.

This movement is not progressive to any significant degree. The bowing of the window jambs takes place over the first few decades of the building's life. Once the load path is able to flow past the narrow jamb brickwork and into the thicker brickwork of the main part of the walls, the movement stops. Similarly, sagging of lintels takes place over the first few years of the life of the building. Once the initial sag has taken place, any continued movement is just minor long-term creep, and this is not significant.

There is however other movement that can occur because of the buckling and bowing in the front wall. As the centre section of the wall bows downwards and is pushed outwards by the roof, it pulls away from the party walls (see Figures 2.18.7 and 2.18.8).

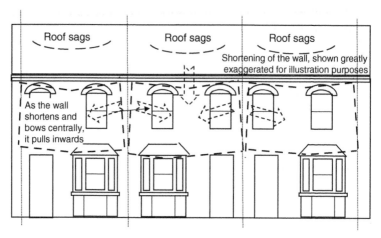

Figure 2.18.7   Distortion of the elevation of the terrace.

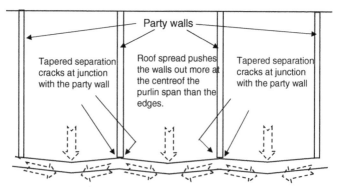

Bowing and shortening of the front elevation pulls the walls in from the party walls. The arrows show the force pulling sideways and away from the party walls.

Figure 2.18.8   Plan view of elevation with bowing from shortening and roof spread.

The dotted lines in Figure 2.18.7 show the distortion of the front elevation from shortening and bowing. This can cause vertical separation between the front elevation and the party walls (or flank wall if it is end terrace). The arrows in Figure 2.18.8 show the direction of the forces pulling the walls inwards, away from the party walls and into the centre of the elevation. Because the wall is bowing, the force is diagonal in respect to the party walls. The diagonal force has both a sideways component and an outwards component. At the party wall positions, the two outwards components combine to pull the front elevation away from the party wall. Cracks can occur in the corners as a result. The cracks will taper slightly, being wider at the top than the bottom giving them the appearance of roof spread. The cracks are likely to be quite small, hairline to 1–2 mm.

Again, please be aware that in practice it is rare to see the separation cracks between the party walls and the elevations because cracks will usually have been covered during redecoration cycles over the lifetime of the building.

In addition to separation of the party walls and elevations, diagonal cracking sometimes occurs in the party walls, or flank wall if it is end terrace (see Figure 2.18.9). This occurs at upper level in the corners of the rooms (and may extend into the roof space). This cracking is caused by the front elevation pulling forward and dragging the party wall with it by friction and the bond of the brickwork.

The cracks are diagonal and get wider at the top. They could be confused with subsidence cracks in appearance but clearly, they are not, because they only effect the upper part of the building. (See Part 3 on movement related to foundations.)

Whether or to what magnitude these separation cracks occur depends not just on the amount of movement, but also on how well the building and brickwork is bonded together. For example, in low specification Artisan housing the party wall might only be half a brick thick. Half brick thickness walls are often referred to as 'four-inch brickwork' because half a brick is four inches (105 mm) thick. Four-inch brickwork in a party wall would be pulled apart more easily than a party wall that was a full brick thick (9 inches, 225 mm).

Again, such movement is usually covered over by redecoration. It is, however, quite common for there to be fitted cupboards/wardrobes in the reveals either side of the chimney breasts. As people often do not decorate inside cupboards, (or don't decorate to the same standard inside cupboards), cracking may remain visible inside the cupboards. Similarly,

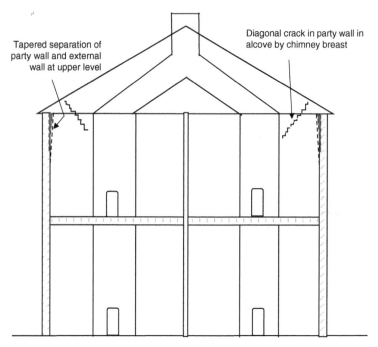

Figure 2.18.9 Section through building showing position and shape of cracks.

it would be very unlikely for cracks in the party walls in the roof space to be filled. If the party walls continue into the roof space, the cracks will also continue into the roof space.

This is quite a complicated series of events that happens slowly over time. It moves as creep, more like the building was made of very stiff plasticene than rigid materials. The movement will start when the house is built and will slowly taper off with time until it halts. It occurs over the first few decades of the buildings life and then stops, for all practical purposes. Whether after 100 years or more, any micro movement is still happening is difficult to say but if it is, it will be at such an insignificant rate that it can be considered as static for all practical purposes.

Consequently, it would be very unlikely that any remedial action would be required. Movement caused by this would not normally be of a magnitude that would require any structural repair. The bowing of the window jambs is likely to be less than 5 mm. The bowing and buckling might not even be visible to the naked eye. You should however be aware that this marginal movement at upper level may also need to be added to movement at lower level. Typically, there will have been some compression at ground floor level, either compression of the sash jambs or sagging of lintels over bay windows. Narrow sections of brickwork between windows will also have been compressed over centuries as a result of the load paths.

Even where this movement is visible it is unlikely to require any remedial work other than for cosmetic reasons.

The description above is based on a house that is symmetrical, front and rear. Old, terraced housing often has a two-storey rear wing, often referred to as an 'out rigger' or rear addition, even though it is original (see Figure 2.18.10). Where there is a two-storey rear section, this acts as a buttress to the rear elevation which restricts (or prevents) the wall from being pushed out. In which case the movement will be restricted to just the front elevation only.

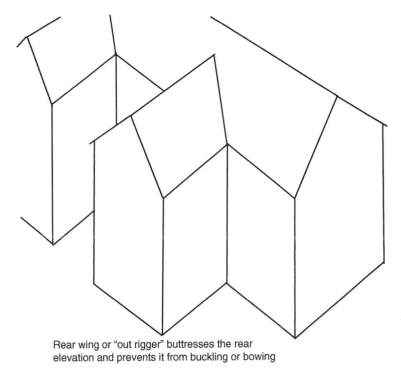

Rear wing or "out rigger" buttresses the rear
elevation and prevents it from buckling or bowing

Figure 2.18.10　Terrace of houses with projecting rear wing or 'out riggers'.

If there is a bay window at ground-floor level, and there has been deflection of the lintel over the window, all of the symptoms are likely to be more pronounced.

Shortening and bowing of the wall will transfer more weight to the roof purlins. All other things being equal this will cause a greater degree of deflection to the roof timbers. Most people assume that deflection in a roof is caused by undersized timbers in relation to the weight of the roof covering. Deflection of the roof caused by this process of wall shortening is generally not understood. If deflection of the roof is seen during an inspection you should follow the trail to see if this is related to wall shortening or just undersized timbers (or both).

Load Path Cracking

Loads are not evenly distributed through walls of buildings. Loads 'flow' down through a building and are diverted by openings in walls. Load concentrations occur at openings where there are beams or lintel supports. Thrust is created at arches. Cracks often occur at the junctions between heavily loaded areas and lightly loaded areas. Traditionally, buildings tended to have narrow openings, aligned on each floor. This form of movement is more likely where openings are wide and not in line on each floor. It is usually minor and insignificant in structural terms.

Key features

1. Small cracks following mortar joints usually under window openings.
2. Cracks form at junctions between loaded brickwork and unloaded brickwork.
3. Usually associated with openings set off line vertically.

There is unfortunately no one single crack pattern for load path cracking. Load path cracking is caused by uneven distribution of the load through a wall. The way the load is distributed is determined largely by the position of openings within a wall.

Practical Guide to Diagnosing Structural Movement in Buildings, Second Edition. Malcolm Holland.
© 2023 John Wiley & Sons Ltd. Published 2023 by John Wiley & Sons Ltd.

If an elevation of a building had no openings at all, the load would be evenly distributed throughout, and no cracking would be likely. Most elevations, however, include openings for windows and doors. The loads have to be transferred or carried around these openings by lintels or arches. This concentrates the loads into certain points. The loads are then channeled down some parts of the walls, almost like columns of load bearing brickwork.

Historically, in most buildings, particularly during the Georgian and Victorian periods, the window openings tended to be vertical in nature and quite narrow. The windows also tended to be in line, from ground floor to upper floors. Not only is this arrangement generally pleasing to the eye, but it is also quite effective from a load path point of view. The narrow openings do not create very large load concentrations at support points. If the windows are all in line, the loads can pass vertically to the foundations.

In more recent decades, in particular during the 1960s to 1970s, window openings became more horizontal and much wider. The architecture also often drifted away from the discipline of keeping windows in line on each floor level.

More recently window sizes have reduced again because of size restrictions imposed by Building Regulations, for energy efficiency. Unfortunately, windows are still often set out of line and therefore, loads have to track across the bond line of the brickwork in order to find their route to the ground.

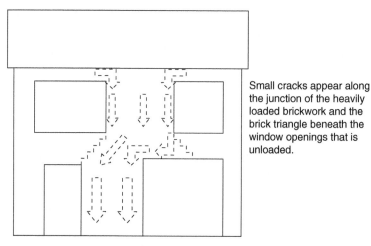

Small cracks appear along the junction of the heavily loaded brickwork and the brick triangle beneath the window openings that is unloaded.

Figure 2.19.1　Load path cracking.

Figure 2.19.1 shows a typical crack pattern that is likely to occur. The loads are diverted around openings into concentrated areas. The path has to divert out of the vertical gravity route to the foundations. The area of brickwork beneath the window openings is unloaded, apart from its own self-weight. With normal daily expansion, seasonal movement and creep over years or decades, a differential movement crack occurs on the load path line. The crack is likely to be hairline or possibly 0.5–1 mm as a maximum.

The cracks are neither structurally significant nor progressive. No remedial action is required other than for aesthetic reasons. Often attempting to re-point the cracks merely serves to highlight them, as the mortar cannot be matched. The result is usually an obvious re-pointed mortar joint, which is less pleasing than just a small crack, which would otherwise be virtually invisible to the eye.

Bulging of Walls Due to Decay of Bonding Timbers

Bonding timbers may have been built into old solid walls for a number of reasons. This was particularly common during the Georgian and Victorian periods. Over time, timbers can deteriorate due to fungal decay (rot), or wood boring beetles (wood worm). This is particularly a risk on exposed elevations. If this occurs, the walls will bow or bulge. This can be a serious defect, if it is not diagnosed at an early stage, and rectified.

Key features

1. Bulging of walls between a full storey height; floor level to floor level.
2. Commonly Victorian or Georgian buildings.
3. Exposed elevations particularly west or southwest facing.
4. Internal crumbling or cracking of plaster on the line of timber built into walls.

Victorian or Georgian buildings sometimes have timbers built into the walls for bonding purposes, or as fixings for panelling. Over the lifetime of a building there is always the possibility of fungal decay (commonly known as rot) or attack by wood boring beetles (commonly known as woodworm). Wood boring beetles find it easier to attack damp timber. Dampness is a pre-requisite for rot. In the United Kingdom, dampness in

---

Practical Guide to Diagnosing Structural Movement in Buildings, Second Edition. Malcolm Holland.
© 2023 John Wiley & Sons Ltd. Published 2023 by John Wiley & Sons Ltd.

walls is more likely to occur on the west and southwest facing elevations, which are the most exposed to the prevailing rain-carrying wind. There could, however, be other causes of dampness such as defective rainwater goods.

As the timber decays it loses its strength. It will physically shrink as it is consumed by the rot or beetles. It will eventually crumble under pressure from the loads of the building. Horizontal cracks will appear on the line of the timber, and the plaster will begin to crumble. The inner face of the wall will be compressed under the load of the building. As the inner face shortens under compression, the wall will bend (see Figure 2.20.1).

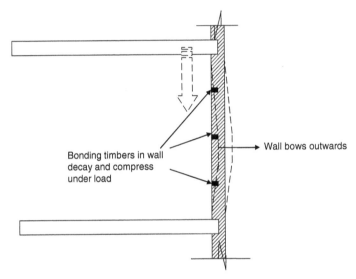

Bonding timbers in wall decay and compress under load

Wall bows outwards

Figure 2.20.1 Bulging caused by decayed bonding timbers.

The extent of the bowing is likely to be more pronounced if the floors run parallel to the wall, as there will be no lateral restraint. (Please also refer to separate notes on lack of lateral restraint.) If the floors are built into the walls at each level, then they will provide some restraint.

As the walls bow and shorten under compression, any floor built into them will go down with them. The floors will move out of level, down towards the bowing wall. If the floors run parallel to the bowing wall, then the floor will remain level.

The extent of remedial work required will depend on how far the wall has moved out of vertical. If movement is minor the timbers can be cut out in sections and replaced with cast in situ concrete. Pronounced movement may require partial or complete rebuilding. This will depend not only on the amount of movement that has occurred but also the intensity of loads on the walls.

Bulging and Separation in Solid Brick Walls

Solid walls are sometimes built as two separate leaves butted together. This was often the case in Georgian buildings, where a single half brick leaf of facing bricks was built to the front of a solid single brick thick wall, built in common bricks. The bricks were often of different sizes and could only be partially bonded together where the brick courses coincided. With time and differential movement, the connecting bricks can crack, and the outer skin separates from the wall. This can be a local defect or can affect a whole elevation or whole building.

Key features

1. Usually 330 mm (brick and a half) thick walls.
2. Bulging not related to floor positions.
3. Commonly Georgian buildings. Flemish or Flemish Garden wall bond.
4. Bulging of the external face, whilst the internal face remains reasonably vertical.
5. More likely on south and west elevations.

When solid brick walls are built as a brick and a half thick (330 mm), they should be bonded together throughout. This is, however, not always the case. Sometimes the walls are not bonded together at every course.

Practical Guide to Diagnosing Structural Movement in Buildings, Second Edition. Malcolm Holland.
© 2023 John Wiley & Sons Ltd. Published 2023 by John Wiley & Sons Ltd.

Historically, the walls were sometimes built in two parts; as a 215 mm thick (single brick) wall, with an outer face of 115 mm thick (half a brick plus mortar joint) brickwork, partially bonded to it.

This was a cost-saving measure, often used during the Georgian period. The inner part of the wall was built in cheap common bricks, at more speed, with less care and using cheaper labour. The outer part of the wall was built using more expensive facing quality bricks, with more care and by the most skilled brick layers.

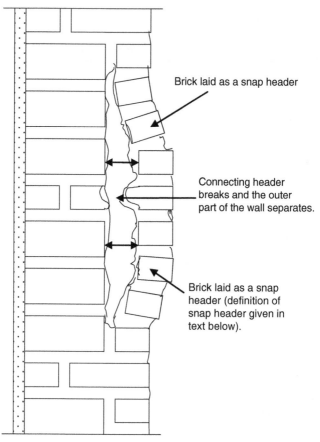

Brick laid as a snap header

Connecting header breaks and the outer part of the wall separates.

Brick laid as a snap header (definition of snap header given in text below).

Figure 2.21.1   Bulging caused by poorly bonded brickwork.

As the size of the common bricks would not match the size of the facing bricks, the courses would not coincide completely. As a result, the outer part of the wall could not be bonded in completely as a homogenous wall. The inner and outer part could only be bonded in where the courses matched or by using bonding timbers in the inner part of the wall, to re-align the joints periodically.

The facing bricks would only be tied in at a few places with complete bricks, laid as headers in the outer leaf, running back into the inner part of the wall. Elsewhere the bricks would be snapped in two and the visible header in the outer part of the wall would not bond into the inner leaf at all. Half bricks laid to show the head (small face) of the brick in the outer leaf, but which do not bond through, are referred to as 'snap headers'.

Bonding timbers may also have been built into the inner part of the wall to allow the fixing of panelling or to bring irregular coursing back into line.

Over the lifetime of the building some movement is bound to occur between the outer part of the wall and the inner part. The outer part is more exposed to the rain and the sun. Seasonal movement will be greater therefore on the south and west elevations.

Timbers built into the wall may shrink and swell seasonally. Timbers in the wall suffer from fungal decay (rot) or attack by wood boring beetles (woodworm). This is more likely on west or southwest elevations in Britain.

The movement between the inner and outer parts of the wall can cause the bonding bricks to snap. The outer part of the wall then detaches from the inner part (Figure 2.21.1). Initially this will occur locally, but it is likely to become progressively worse with time.

Provided that the inner part of the wall remains plumb and stable, then remedial work involves re-building just the external bulged areas. The rebuilt areas are connected to the original brickwork with stainless steel ties.

2.22

Separation of Rubble-Filled Stone Walls

Stone walls are often built as an inner and outer leaf of stone, with a rubble-filled core. The inner and outer stone leaves are tied together periodically, with bonding stones through the full thickness of the wall. The rubble-filled core can sometimes consolidate with time. Voids may form in the wall. If the wall is poorly bonded, it may separate or bulge. This can be a localized defect requiring minor repairs or can be so severe that re-building is required.

Before machinery was available to cut and dress stone, it had to be cut and dressed by hand. This is hard and laborious work. Consequently, the stone would usually be sorted into roughly evenly sized pieces, so that it could be laid in approximate courses, without much dressing.

Most old stone walls are built with an inner and outer leaf aligned reasonably regularly externally on both faces, with all the uneven bits facing into the core of the wall. The core is then filled with any bits that have been cut off to dress the stone, plus any other rubble debris. This may also include mortar and clay (Figure 2.22.1).

To hold the inner and outer leaf together, and stop the wall from bulging or becoming unstable, large cross stones or 'bonding stones' are periodically laid through the entire width of the wall.

Practical Guide to Diagnosing Structural Movement in Buildings, Second Edition. Malcolm Holland.
© 2023 John Wiley & Sons Ltd. Published 2023 by John Wiley & Sons Ltd.

Stone walls cannot be laid as straight or plumb as brick walls and the same regular bond cannot be achieved. As a result, stone walls are not as strong or stable as brick walls, given the same thickness. Consequently, stone walls have to be built much thicker than brick walls, commonly 450–600 mm.

Old stone walls are often rubble filled with an inner and outer leaf of stone. Bonding stones are laid across the wall every few courses. The centre of the wall is rubble, mortar and/or clay. Consolidation of the fill can create voids. If insufficient bonding is provided during construction, the inner and outer leafs can separate and bulge.

To strengthen old walls, holes are drilled into the centre of a rubble core and grout is introduced, to fill any voids and stabilize the wall.

Figure 2.22.1   Rubble-filled stone wall construction.

If the fill was not well compacted at the time of construction it will consolidate with time, under gravity. Voids may also form within the wall. If the wall is not very well cross bonded, this can cause the outer leaf to separate and bulge.

If movement is minor and detected at an early stage, the walls can be stabilized by filling the wall with a slurry or grout.

If the outer leaf of the stone has bulged to any significant degree, the affected areas will have to be cut out and rebuilt. Where large areas are affected, buttresses or tie bars may also be required. In the worst-case scenario demolition and re-building may be required.

2.23

Floor Slab Settlement (Compaction)

Ground bearing concrete floor slabs are laid on hardcore. If the hardcore was not properly compacted at the time of construction, it will settle naturally with time. This can cause the floor slab to settle and drop. Similarly, if the hardcore was contaminated with clay when it was laid, settlement can also occur, as the clay shrinks after construction. The movement and cracking associated with this defect can vary greatly. The pattern of movement depends on whether the internal walls are built off the slab or whether they have independent foundations. This defect can be localized and minor, or widespread and severe. In minor cases repairs may involve just the application of a levelling topping. In severe cases the floor may need to be underpinned, or completely excavated and re-laid.

Floor slab settlement (often referred to as floor slab compaction) is caused when the hardcore fill beneath a concrete ground bearing floor slab compacts with time. This is usually because the hardcore was not sufficiently well compacted when it was first laid.

Solid concrete, ground bearing floors were not widely used until after World War Two. Before World War Two, most floors were suspended timber. Where solid floors were used before this time, mostly in kitchens, they would be typically quarry tile, laid on an ash, sand or mortar bed. Concrete was quite rare and expensive before World War Two.

Practical Guide to Diagnosing Structural Movement in Buildings, Second Edition. Malcolm Holland.
© 2023 John Wiley & Sons Ltd. Published 2023 by John Wiley & Sons Ltd.

There was, therefore, little experience of this type of floor amongst the labour force, until some years after World War Two.

The construction process is as follows (Figures 2.23.1 to 2.23.3):

Stage 1. The topsoil is stripped and foundation trenches are dug. The strip footings are cast in the trenches and the brick walls are built up to damp proof course level.

Figure 2.23.1   Ground bearing floor slab construction process stage 1.

Stage 2. The trench is backfilled externally. Hardcore is laid internally to fill the trench and a bed of hardcore is laid to form a level bed on which to lay the concrete. The hardcore should be laid in beds not exceeding 300 mm in depth and well compacted at each bed level to bring it up to the required level.

Figure 2.23.2   Ground bearing floor slab construction process stage 2.

Stage 3. The hardcore is surfaced with sand and a plastic damp proof layer is laid on it. The concrete is poured on top using the external walls to contain it. The slab forms a 'raft' which 'floats' on the hardcore. The hardcore transfers the loads from the slab to the ground below, evenly over every square millimetre. Hence the terminology 'ground bearing slab'.

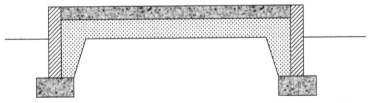

Figure 2.23.3   Ground bearing floor slab construction process stage 3.

Problems occur if the hardcore is not compacted firmly at construction. If voids are left during compaction, it will settle down under gravity, over months or years after construction. The deeper the fill under the slab, the greater the risk will be of incomplete compaction at construction. If the site is sloping and there is more fill at one end than the other, the greater is the risk of uneven compaction. The risk tends to be greater in corners where it is more difficult to mechanically compact.

The maximum recommended depth of fill under a slab today is 600 mm. Where the required depth of fill to achieve a level would exceed 600 mm, a suspended timber or suspended concrete floor should be used. Historically, this recommendation did not exist and the depth of fill often exceeded 600 mm. On sloping sites one corner or one edge may be considerably above 600 mm.

Poor compaction is the main cause of floor slab compaction, but contamination of the hardcore is also sometimes a factor. If the hardcore is contaminated by soil, particularly clay, this will shrink as it dries out after construction and settlement will occur.

It is difficult to describe one particular crack pattern associated with this type of movement as the compaction may be localized or widespread.

The cracking and movement is also different depending on whether the internal walls are built off the slab (commonly the case in bungalows) or built with independent footings.

If the walls are built off the slab, the cracking may also vary depending on whether the roof is supported on the walls or not.

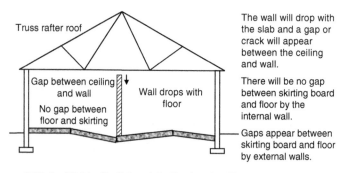

Figure 2.23.4   Wall built off the slab. Roof and ceiling not supported on internal walls.

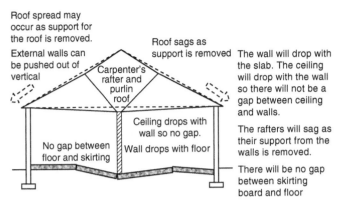

Roof spread may occur as support for the roof is removed. External walls can be pushed out of vertical

Carpenter's rafter and purlin roof

Roof sags as support is removed

No gap between floor and skirting

Ceiling drops with wall so no gap.

Wall drops with floor

The wall will drop with the slab. The ceiling will drop with the wall so there will not be a gap between ceiling and walls.

The rafters will sag as their support from the walls is removed.

There will be no gap between skirting board and floor

Figure 2.23.5 Wall built off the slab. Roof and ceiling supported on internal walls.

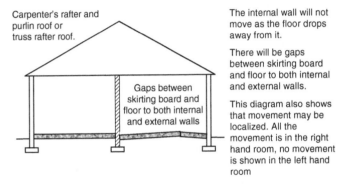

Carpenter's rafter and purlin roof or truss rafter roof.

Gaps between skirting board and floor to both internal and external walls

The internal wall will not move as the floor drops away from it.

There will be gaps between skirting board and floor to both internal and external walls.

This diagram also shows that movement may be localized. All the movement is in the right hand room, no movement is shown in the left hand room

Figure 2.23.6 Wall built with independent foundation.

Figure 2.23.4 shows a common scenario in bungalows built from the late 1960s onwards, when truss rafter roofs were introduced. The introduction of truss rafters made it possible to span a roof from external wall to external wall, without any support from internal walls. The internal walls merely divide up the spaces in this form of construction. Blockwork walls were usually used for their sound proofing quality. The same could also be achieved with a TRADA truss roof, commonly found from after World War Two, up until truss rafters were introduced.

As the internal walls are not load bearing, they do not need foundations and are usually built off the floor slab. When floor slab compaction occurs and the slab subsides, the internal walls will go down with it. As a result, the wall will pull away from the ceiling that is fixed to the underside of the rafters and a crack or gaps will occur.

The skirting boards will be fixed to the internal wall, and they will therefore go down with it. As a result, there will not be a gap between the floor slab and the skirting fixed to the internal wall.

The external walls, however, will always have independent foundations. If the slab subsides next to the external walls, the floor will drop away from the walls. A gap will therefore appear between the floor and the skirting boards where the skirting boards are fixed to the external walls.

The second diagram, Figure 2.23.5, shows a bungalow built with a carpenter's roof. A carpenter's roof is made up with rafters cut on site from timber. Intermediate support is provided by horizontal timbers (purlins) at right angles to the rafters. This type of roof was the universal method of construction before World War Two. It was still commonly used after World War Two but became increasingly superseded by TRADA truss roofs between World War Two and the late 1960s. From around the late 1960s truss rafters were introduced and they have progressively taken over as the most common, almost universal, method of roof construction in the United Kingdom.

Rafters and purlins are only able to achieve fairly small spans, commonly around 2.5–3 m. Support has to be provided to the purlins by struts down to internal walls. This causes load concentration points at the support points. If internal walls are used to support a roof, they should have independent foundations, but this is not always the case. If the walls are built off the slab, the degree of movement is likely to be greater at the points of load concentration. That is, of course, subject to all other things being equal.

When the floor slab settles, the walls will go with the floor. If the walls support the roof by struts, then the roof will go down at the support points. Removal of the support will cause the purlins to sag, and in turn, the rafters will sag. This will sometimes allow the ridge of the roof to drop and the rafters to push outwards causing roof spread. The external walls will be pushed out of vertical as a result, and tapered cracking may be recorded at the junctions between internal walls and external walls. (Please refer to previous notes on Roof Spread in Chapter 2.5.)

As the ceiling comes down with the walls, there will not be separation between the ceiling and the walls.

As the internal walls go down with the floor, there will not be gaps between the internal walls and the skirting boards. The external walls, however, will always have independent foundations. If the slab subsides next to the external walls, the floor will drop away from the walls. A gap will therefore appear between the skirting and the external wall.

The third diagram, Figure 2.23.6, shows a scenario where the internal walls have independent foundations. Load bearing walls would normally have independent foundations. In a bungalow, load bearing walls would commonly be used where they support the roof or ceilings. In a house, the ground-floor walls usually support the floors and at least some of these are likely to be load bearing.

When the floor settles, the walls with independent foundations will not move. As a result of which gaps will occur between the skirting boards fixed to the internal walls and the surface of the floor.

The ceilings will either be clear spanning and fixed to the underside of a trussed roof or will be fixed to joists and supported by the load bearing walls with independent foundations. The ceilings and walls will therefore, not separate and there will be no cracking or gaps at ceiling level.

In all cases there will be some common factors, and this will vary by degree depending on the extent of movement.

The amount that a floor will subside depends on where and to what degree the hardcore compacts. Movement may be isolated, and this is commonly the case on sloping sites where it is only the deeply filled areas that compact with time. If poor workmanship creates widespread, poorly compacted hardcore at the date of construction, then movement may be general and widespread. It may also, of course, be a mixture of the two.

The size of the gaps beneath the skirting boards may therefore vary but they are commonly 10 mm or even more. The gaps may be even in width but may also taper if the movement is uneven. In most cases the gaps will get wider in the corners where the movement is usually greatest.

Concrete is a hard, brittle material and therefore when the slab settles unevenly, it will crack. Changes in floor level will occur because of uneven movement.

If water pipes are built into the slab, these may break and leak. When the slab settles it will often pull the damp proof membrane out of the wall. Perimeter dampness can occur which, although usually fairly low, can attract insects particularly woodlice.

A concrete floor slab has to be laid into the reveal of door openings to external walls. When the floor settles it cracks across the door reveal as the section of slab over the external wall is supported by the wall beneath. Similarly, if an internal wall has independent foundations, cracking will occur across door openings.

Remedial action for this type of movement is dependent on the degree of movement in each case. If the movement is only minor and isolated, possibly no remedial work will be required. In other cases, it might be possible to simply re-level the floor surface using a self levelling screed or traditionally an asphalt screed. An asphalt screed also has the benefit of being waterproof.

If re-levelling requires anything other than a very thin self-levelling screed, it will be necessary to re-fit all the skirting boards and re-hang or replace the doors. Kitchen units will also probably need taking out and re-fitting. Re-fitting the associated items can make even relatively moderate movement expensive to repair.

If any pipework is set into the floor, movement will probably rupture the joints. Re-installing pipes for water and central heating is another cost item.

In even more severe cases, the floor may need to be underpinned. There are a number of methods available. Sometimes a grout can be pumped in under the floors to stabilize them. Floors can be injected beneath with expanding foam to lift them back into place. The floor can be drilled for the insertion of mini piles. Alternatively, in the most severe cases, the floor may have to be completely excavated and re-laid.

The magnitude of repairs is dependent on the facts in each case but can be very expensive. As this is usually a defect attributed to workmanship at the time of construction, it is usually viewed as an inherent defect and may not be covered by buildings insurance.

It has the potential to be about the most expensive defect to repair that is not covered by buildings insurance.

2.24

Load Concentrations

Buildings are not uniformly loaded. The more heavily loaded parts will be compressed or will settle more than the light loaded areas. This causes some deflection or sagging with age. Provided that this compression or settlement is not too severe, no remedial action is required.

Most buildings are designed to have reasonably evenly distributed loads. However, load concentration points will occur for a number of reasons. Wherever there are large openings in walls, the weight above the opening has to be carried by a beam.

Load concentrations occur at the support points. If the loads in a building are not evenly distributed, one part of a wall will be loaded more heavily than another part.

With time, over years and decades usually, the heavily loaded part will settle into the ground. As it does so, the ground will become compressed. The supporting wall itself will also be compressed more where it is heavily loaded, as a consequence of the natural elasticity of the material itself.

Load concentrations are also caused internally, by storage or by water containers, for example, water tanks, hot water cylinders and baths.

Load concentrations are also found beneath roof support points where struts are taken from the roof timbers down to load bearing walls. It is

Practical Guide to Diagnosing Structural Movement in Buildings, Second Edition.
Malcolm Holland.
© 2023 John Wiley & Sons Ltd. Published 2023 by John Wiley & Sons Ltd.

also quite common for roofs to be strutted to partition walls at first-floor level that were never intended to be load bearing. If the partitions are strong enough to carry the loads, they can transfer them to the floors. This often causes floors to sag.

It is therefore essential to be able to sketch a building and work out how the loads flow through the walls and where the concentrations occur. Load distribution has already been described in Chapter 1.6.

Once the flow of the loads is determined, the heavily loaded parts can be identified. Provided that a building has only sagged where it is meant to sag (in the heavily loaded areas) and the magnitude is not too severe, no remedial action is necessary.

Sulphate Attack

Sulphate attack is a chemical reaction that takes place between sulphates and cement in the presence of water. Cement contains tri calcium aluminate which reacts with sulphates from the ground or building materials, such as bricks or plaster. The reaction causes expansion and cracking. External renders are particularly vulnerable to sulphate attack on exposed elevations. Sulphate contamination in hardcore, from old gypsum plaster or furnace waste, can lead to sulphate attack in concrete floors. It can be a cosmetic defect or a serious defect. The severity of the defect will be dependent on the facts in each case. It can range from purely cosmetic, to a significant structural problem.

Key features

1. Random cracking in renders.
2. Concentration of cracking in wet areas of walls, for example, under window sills.
3. Cracking on lines of unlined chimney flues.
4. Spalling of brickwork where set in cement mortar.
5. White staining in mortar joints running out into cracks in the bricks.
6. Bulging and cracking of concrete floors.
7. Undersail at damp proof course level associated with floor expansion.

Practical Guide to Diagnosing Structural Movement in Buildings, Second Edition.
Malcolm Holland.
© 2023 John Wiley & Sons Ltd. Published 2023 by John Wiley & Sons Ltd.

Sulphate attack is a chemical reaction. In order for the chemical reaction to take place these factors are required:

1. The presence of sulphates.
2. Ordinary Portland cement.
3. Water.

In addition, a fourth factor is required which is time, as it is a slow reaction.

Sulphates react with tri calcium aluminate, found in ordinary Portland cement. This reaction causes expansion, leading to cracking.

It should therefore be noted that sulphate attack will only occur where modern cement has been used in a mix of mortar, render or concrete, etc. Older properties, where the mortar or render is lime based, cannot be subject to sulphate attack because there is no cement. If there is no cement, there is no tri calcium aluminate and no sulphate reaction.

It is a common misconception that the sulphates come from within the cement. This is not the case. The sulphates must come from elsewhere. This can be from, for example, bricks, hardcore, plaster or the ground.

Probably the most common example of sulphate attack is found in external renders. If an exposed elevation of a solid brickwork building is suffering from penetrating dampness, it is quite common for a render to be applied in order to try and prevent this. Traditionally, when a soft lime-based render was used, this may have worked. Unfortunately, most builders and probably even some surveyors would now use a cement: sand render. It is easy to understand why someone would think that using a very hard cement rich mix would be more effective than a weak mix.

In fact this approach is flawed. A cement-based render will contain tri calcium aluminate. The clay bricks, to which the render has been applied, will contain sulphates. When cement-based render 'goes off', there is a chemical reaction that gives off heat. This causes expansion, followed by shrinkage after the chemical reaction has taken place. This results in invisible cracks in the render. The richer the mix, the more cracks there will be.

When rain hits the wall, it will be sucked through the cracks by capillary action. Once behind the render, it will not be able to evaporate away because the render restricts evaporation from the surface. The result is entrapped dampness in the wall. The water dissolves sulphates from the bricks into solution and brings them together with the tri calcium aluminate in the cement.

The result is random crazed cracking in the render (Figure 2.25.1 and Figure 2.25.2). If the wall contains a chimney, there is likely to be a higher concentration of sulphates from decades of combustion products soaked into the brickwork around the flue. Cracking is then likely to be greater following the line of the flue (Figure 2.25.1). As water is required for the reaction it will be likely to occur on the exposed elevations, normally the west and southwest in the United Kingdom.

Once the render has cracked, it will allow more moisture to penetrate the surface and the process will accelerate. This cracking is not structural in nature. One solution would be to remove the render and provide an alternative method of weather screening. Alternatively, a weather screen cladding could be fixed over the render, thereby excluding water and preventing any further chemical reaction.

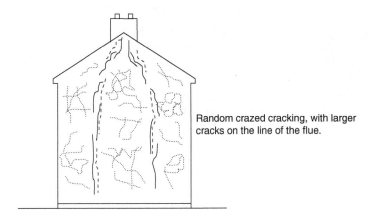

Random crazed cracking, with larger cracks on the line of the flue.

Figure 2.25.1 Sulphate attack in render.

Sulphate attack may also occur in the mortar joints of brickwork. In all modern clay brickwork, there will be cement in the mortar and sulphates in the bricks. Although the two are found together, sulphate attack is rarely seen because the walls do not usually remain wet for a sufficiently long period of time.

Sulphate attack is seen quite commonly in free-standing boundary walls, which do remain wet for long periods. It is quite common to see brick on edge copings in garden walls. The bricks are often cracking and spalling (flaking) at the top. This is usually attributed to frost damage. It is often caused, at least to some degree, by sulphate attack. The coping bricks are usually laid in a strong mortar mix of 1:3 cement: sand. The bricks are often not as strong as the mortar. As the sulphate attack causes expansion of the mortar joints, the bricks get crushed. To differentiate between sulphate attack and frost damage in this situation, look for white staining in the cracks. White staining and whitening of the mortar are a

Figure 2.25.2　Random cracking in render on exposed Westerly elevation.

common features of sulphate attack. In most exposed garden walls, it will be both a combination of sulphate attack and frost damage acting together. For practical purposes the consequence is the same, the bricks begin to crack and spall.

Another example where sulphate attack is found is in concrete floors, in particular garage floors. In a garage floor it is possible that there is no damp-proof membrane to separate the concrete from the hardcore. Broken bricks often used as hardcore will contain some sulphates, but the most severe case is where furnace waste is used. Furnace waste contains large amounts of sulphates.

Another common source of sulphates is recycled rubble containing gypsum plaster (gypsum plaster is calcium sulphate).

If the hardcore contains sulphates, it can react with the concrete causing it to expand. As the concrete expands it will push the walls outwards. If the walls are not heavy enough to resist the outwards thrust, the brickwork will project or 'under sail' at floor level (Figure 2.25.3 and Figure 2.25.4). If the walls are heavy enough to resist the outwards thrust, the floor will bow upwards. In practice, both may occur at the same time.

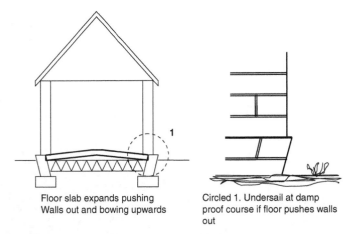

Floor slab expands pushing
Walls out and bowing upwards

Circled 1. Undersail at damp
proof course if floor pushes walls
out

Figure 2.25.3 Sulphate attack in concrete floor.

This can be structurally significant. To rectify this problem, it is necessary to remove one of the three factors required for the chemical reaction to take place. One way to achieve this would be to excavate the floor and relay it with new hardcore that did not contain sulphates.

Alternatively, remove the concrete, lay a plastic membrane on top of the old hardcore and recast a new concrete slab. The membrane would separate the concrete from the sulphates in the hardcore. It would also provide a damp proof layer, thereby removing water from the equation and preventing further reaction.

Sulphate-resisting cement can also be used in render, mortar or concrete mixes. It should be noted that this cement is only sulphate resisting, not sulphate immune. These cements contain less tri calcium aluminate and therefore any reaction is less aggressive.

After World War Two, and up until relatively recently, it was quite common practice to plaster houses internally with 11 mm thick cement sand base coat, skimmed with a 2 mm thick gypsum finish coat. Gypsum is calcium sulphate and is incompatible with a cement sand base coat if moisture is present.

It is quite common for the plaster finish coat (skim coat) to detach from the base coat after a number of decades. This is usually attributed to the possibility of cold weather during plastering, dirt on the base coat or poor 'keying' of the base coat. 'Keying' is scratching the base coat to form grooves into which the skim coat can stick. It is possible that what is

Figure 2.25.4   Under sailing of brickwork below damp proof course where expansion of the floor slab has pushed the brickwork out.

actually happening is that moisture is being picked up very slowly as vapour from normal activity in the building. Over a period of decades, this might cause a sulphate reaction between the cement sand base coat and the gypsum finishing coat, causing it to separate.

As previously mentioned, modern brickwork does not usually suffer from sulphate attack even though sulphates and cement are present together because the walls do not remain sufficiently wet for long periods of time.

Now that cavity walls are commonly filled with insulation, the external leaf will not be dried out as quickly by heat escaping from the house, as was previously the case. As a result, the brickwork will remain wetter for longer. In theory, this could increase the risk of sulphate attack in all modern brickwork. To date, it does not appear to be having any significant effect, but over decades, there is the possibility that it might. Some elements of a building, such as the walls, are usually expected to last decades or centuries and this gives a lot of time for the reaction to take place. Building defects often take years or decades to develop and only time will tell.

2.26

Concrete Block Shrinkage

Concrete blocks shrink soon after manufacture. This shrinkage causes cracks to occur soon after construction. If the blocks become wet during construction, the cracks will be larger. Cracks in the blocks may also be exacerbated by the use of very strong mortar in the blockwork. Modern foamed cement blocks suffer more cracking than heavy aggregate concrete blocks. The cracks can be large. They can look serious because of their size but they are not structurally significant.

Key features

1. Concrete blocks, steam slaked.
2. Aggregate blocks and modern light-weight 'foamed' cement blocks.
3. Straight vertical cracks of even width.
4. Usually, cracks are hairline width to 2 mm but can be up to 5 mm.
5. Not progressive, cosmetic repairs only.

Shrinkage of concrete blockwork and modern aircrete blocks is likely to be the one of the most common and possibly the most severe looking non-structural cracking that will be seen. Most commonly the cracks will vary in width from hairline to about 2 mm, but it is not uncommon to find cracks as wide as 3 mm, 4 mm or even 5 mm.

Practical Guide to Diagnosing Structural Movement in Buildings, Second Edition.
Malcolm Holland.
© 2023 John Wiley & Sons Ltd. Published 2023 by John Wiley & Sons Ltd.

Originally, concrete blocks (often referred to generically by laymen as 'breeze blocks') were made from aggregate, mixed with cement as the binder. Modern aircrete blocks are cement slurry foamed by a chemical process to form a block of cement, full of bubbles for insulation purposes.

The blocks are steam cured during production and absorb moisture in the process. After manufacture, the blocks dry out and undergo an initial shrinkage. The blocks are also relatively absorbent and, if they get wet during construction, they will expand. The blocks will shrink again as the building dries out after completion. The cement in the blocks continues to hydrate for years after production. This hydration process causes the blocks to shrink a little in the long term.

The magnitude of any shrinkage will depend on the characteristics of the aggregate. Generally speaking, aggregate blocks are unlikely to suffer from shrinkage cracking to the same degree as light-weight aircrete blocks. This is because the aggregate forming the block is unlikely to be as absorbent as the 'aircrete'. Also, there is less cement content in the aggregate blocks, resulting in less hydration later on in service. This is, however, a bit of a sweeping statement. If the aggregate is moisture absorbent and the blocks become wet during construction, the drying shrinkage can be just as pronounced.

Cracking in aircrete blockwork walls is often exacerbated by the blockwork having been built with the wrong strength of mortar. Aircrete blockwork walls should be laid using a soft, weak mix mortar, typically no stronger than, 1:5 cement: sand or 1:1:6, cement: lime: sand.

If a stronger mortar is used, which it commonly is, then the mortar will be stronger than the blockwork. If the cement is stronger than the blocks, any movement will crack the blocks to a greater degree than the mortar. Cement-rich mortar will also shrink more than a weak mortar mix, increasing the risk and magnitude of cracking.

The appearance of the cracking is usually quite vertical. It follows the weakest route through the line of the perpendicular mortar joint and then continues vertically, through the centre of the block in the course of blockwork above. Alternatively, the vertical path will go through the blocks at the lap of the bond. It may, of course, do a little of both, veering off from being completely vertical but remaining mostly vertical overall.

Cracking can occur anywhere in a building, but it will tend to find weak routes, for example, through blockwork beneath window openings. This is also one of the most likely places for radiators to be positioned.

After construction, the walls dry out more quickly next to where the radiators are positioned, which also increases the risk of cracking.

With time, cracks in living accommodation are often filled or covered with wallpaper, and they become hidden. In cupboards, or other similar places, cracks are often left unattended. In garages and roof spaces the blocks are left bare, and cracks will again, usually be ignored for decades or forever. Large blockwork shrinkage cracks are often found in such areas (see Figures 2.26.2 and 2.26.3).

The cracks themselves are even in width and all of the displacement is in the horizontal plane (Figure 2.26.1).

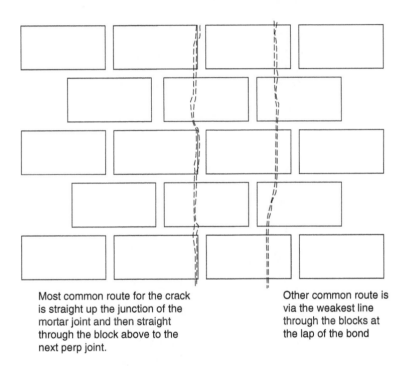

Most common route for the crack is straight up the junction of the mortar joint and then straight through the block above to the next perp joint.

Other common route is via the weakest line through the blocks at the lap of the bond

Figure 2.26.1 Concrete blockwork shrinkage.

Figure 2.26.2   Blockwork shrinkage crack. Note the crack is slightly wider at the top where there is no load.

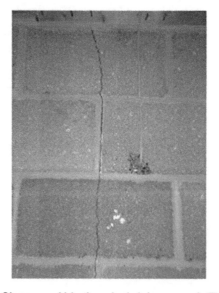

Figure 2.26.3   Close-up of blockwork shrinkage crack. The crack is 2–3 mm wide at top.

2.27

Shrinkage of Calcium Silicate Bricks

Most bricks are clay but some are calcium silicate (sand lime). Calcium silicate bricks are not fired like clay. They are steam cured and hardened by a chemical reaction. A lot of moisture is absorbed during the process. As the bricks dry out they shrink. This is irreversible. The bricks have a high coefficient of expansion and are also prone to expansion cracking. Bricks are often yellow, pink, bluey grey colour. These bricks are sometimes used as an alternative to clay particularly during construction boom periods when there can be a shortage of clay bricks.

Key features

1. Calcium silicate bricks.
2. Initially random cracking, not orientation specific.
3. Secondary expansion cracking more on Southern elevations.
4. Cracking takes weak routes through openings.
5. Horizontal displacement only.
6. Cracks commonly hairline to 2 mm horizontally and hairline on bed joints.
7. Cracks in mortar joints.
8. Cracks in external leaf only.
9. Bricks are often pale yellow, pale pink, blue grey in colour and smooth faced.
10. From 1980s often moulded faced and sand faced to look more like clay bricks.

Practical Guide to Diagnosing Structural Movement in Buildings, Second Edition.
Malcolm Holland.
© 2023 John Wiley & Sons Ltd. Published 2023 by John Wiley & Sons Ltd.

Most bricks in the United Kingdom are made from fired clay. Some bricks, however, are made from calcium silicate (sand: lime). These bricks are not fired in a kiln. They are steam cured (slaked). The hardening process is a chemical reaction.

A lot of moisture is absorbed during this process. After manufacture the bricks dry out and shrink to some degree. This initial shrinkage is not reversible. If the bricks are laid before the shrinkage has taken place, quite a lot of cracking commonly occurs.

The bricks are relatively strong, so the cracking usually occurs in the mortar joints rather than through the bricks. The cracking is likely to look quite random and will take weak routes through narrow sections of brickwork between window and door openings.

The displacement is horizontal, like expansion cracking. Gravity will naturally close up any shrinkage on the bed joints. It is only the effect of sliding and dragging of the bricks which will cause hairline separation on the bed joints.

Cracks are likely to be hairline, up to about 2 mm. The degree of cracking will be worse if a hard mortar has been used. Unless calcium silicate bricks have been used to build the inner leaf of a wall, which would be unlikely, then cracking will be restricted to the outer leaf containing the calcium silicate bricks. Although be aware that blockwork shrinkage in the inner leaf, may also cause cracks in the corresponding position internally. See Chapter 2.26.2 Concrete Block Shrinkage.

Apart from in a few specialist situations, calcium silicate bricks are usually considered as a second choice in comparison to clay bricks. During building boom periods, there is often a shortage of clay bricks. Under such circumstances, calcium silicate bricks are often used as an alternative. Calcium silicate bricks can be made to look very much like sand-faced clay bricks, but before the 1980s the vast majority were not disguised in this way.

They are formed in a mould and therefore the appearance of the face of the brick depends on the mould. The vast majority are smooth faced. Where a textured mould has been used to create a textured face, this tends to look very much more 'plastic' in appearance than does a fired brick. The texture is of a much regular repeated texture, rather than the random texture of heat-fired bricks. This is simply because there are a limited number of mould patterns used during manufacturing.

Prior to the 1980s, the bricks would usually be smooth faced and often pale pink, pale yellow or blue grey in colour.

From 1980s they may have been cast in a textured mould, to imitate a textured brick, colouring may also have been added. Again, the colouring is more consistent than the variable colouring of a clay brick, which relies on the colour of the clay.

It can sometimes be difficult to tell the difference between a sand-faced calcium silicate brick and a sand-faced clay brick. Look for bricks that have been slightly damaged or chipped whilst they were being laid. If the sand face of a clay brick has been chipped off, the clay-fired colour will be exposed behind. Calcium silicate bricks, however, will have a sandy texture that will run through the brick. Often, larger grains, like little stones of a millimetre or two across, can be seen in the matrix of the brick. Imagine the variation in grain size of sand on a beach.

An added complication with calcium silicate bricks is that they have a much larger coefficient of expansion than clay bricks.

If after the initial shrinkage, the cracks are re-pointed, they are likely to recur seasonally due to expansion and contraction. This is particularly the case on south- and southwest-facing elevations.

If the cracks are not pointed up after the initial shrinkage, they will act as natural expansion joints.

If these cracks have formed diagonally, through weak routes between window openings, it can distort the overall appearance of later expansion cracks making diagnosis more difficult.

Please refer to Chapter 2.1 Expansion Cracking.

2.28

Heat Expansion of Flue Blocks

Concrete flue blocks are sometimes built into inner leaves of cavity walls to create a flue for gas fires. Expansion of the flue causes cracking at the junction between the flue blocks and the normal wall blocks. Cracks can be a couple of millimetres. Cracking is not structurally significant but difficult to repair as cracking recurs with each use of the flue as it heats up. Typically found in buildings 1960s onwards.

Key features

1. Concrete flue blocks built into inner leaf of cavity walls.
2. Cracks vertically or staggered at the junction between the flue and the normal wall joints.
3. Cracks even width and typically 1−2 mm wide, possibly 3 mm in some cases.

Prior to about the 1960s, virtually all houses were designed and built to be heated by solid fuel open fires. Virtually all houses had to be built with fireplaces and chimney stacks.

From around the 1960s onwards, open fires began to be replaced by gas fires and eventually by gas-fired central heating boilers and hot water radiators.

Practical Guide to Diagnosing Structural Movement in Buildings, Second Edition. Malcolm Holland.
© 2023 John Wiley & Sons Ltd. Published 2023 by John Wiley & Sons Ltd.

Gas fires and boilers still require a flue to exhaust the products of combustion. The flue requirements are, however, much less complicated than a chimney for an open fire.

A simple solution for the provision of a flue for a gas fire was to build it into the internal leaf of a cavity wall using a hollow concrete flue block. The hollow core blocks are the same width as a typical concrete block (typically 100 mm wide × 440 mm long × 215 mm high) (see Figure 2.28.1). They are usually designed with the flue off set from the centre, so that they can be staggered and built into an inner leaf, thereby maintaining the bond of the brickwork. Although they are most commonly staggered into the bond, this is not always the case. Some blocks have the flue centrally and they stack up with a vertical joint to the normal blockwork at each side (see Figure 2.28.2).

Precast concrete flue block with the flue void set off centre.

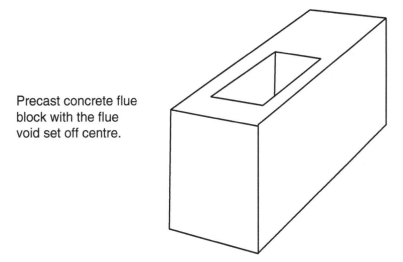

Figure 2.28.1   Example of a flue block.

The flue blocks are usually built up to the roof space and then connected by a flue pipe to a vented ridge tile. Sometimes, particularly when the flue is in a flank wall, the flue blocks are built right up to a chimney stack, corbelled from the wall (see Figures 2.28.2 and 2.28.3).

Flue from flue blocks connects to a flue pipe in the loft which in turn connects to a vented ridge tile.

Flue blocks connects to a brick chimney stack corbelled out from the gable wall.

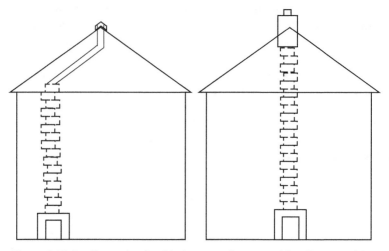

Figure 2.28.2   Alternative flue block connections, to a ridge tile or to a false chimney.

The precast flue blocks are made of a dense concrete. The blocks tend to be denser than the concrete blocks that make up the inner leaf of the external wall, particularly in more modern houses where the inner leaf is a light-weight 'aircrete' block, designed for thermal insulation. (See Chapter 2.26 Concrete Block Shrinkage.)

Heat from the flue gases causes the blocks to expand and contract. This often results in cracking from heat expansion and differential movement between the flue blocks and the inner leaf blocks.

The cracking will most commonly be a stepped in and out crack following the bond of the precast flue blocks with the ordinary inner leaf blocks or if the flue blocks are not bonded, a straight crack on the line of the joint. In some instances, it can be a straight vertical crack in the plaster on the line of the flue. It may also be a combination of the above. The cracks are likely to be hairline to 1–2 mm but possibly as wide as 3 mm.

As the movement is caused by heat expansion and differential movement, it is a cosmetic defect only, and will not get progressively worse with time. It is, however, an unsightly internal crack, usually in the living room and bedroom above. Owners do not want such cracks in such visible locations.

Figure 2.28.3 Flue blocks built into the inner leaf of the gable wall.

The problem with this type of crack is how to repair it. If the crack is simply filled or re-plastered, it will crack again as and when the fire is used. Wallpaper might be flexible enough to stretch over the crack. This is not always the case; it is common to see ripped wallpaper on the line of the flue. One possible option is to fix timber battens to the wall and line the area with plasterboard to create a false chimney breast over the cracked area.

2.29

Floor Cracking in Suspended Concrete Beam and Block Floors

Historically ground-floor construction in the United Kingdom was suspended timber or solid. In response to a number of technical and economic factors the use of suspended concrete floors became more common during the 1980s. By 1990 onwards suspended concrete floors became the most common form of ground floor construction. Inverted 'T' shaped beams span from wall to wall with a vented void beneath. Suspended floors can be identified by air bricks at the base of the wall. At junctions of floor spans, commonly at door thresholds, cracking often occurs.

Key features

1. Buildings from 1980s onwards and particularly 1990s onwards.
2. Suspended beam and block floors.
3. Identify by age of the building and air bricks to the base of the wall.
4. Cracking across internal door thresholds and sometimes external door thresholds.
5. Not significant structurally.
6. Not progressive but can recur.
7. Cosmetic repair.

Practical Guide to Diagnosing Structural Movement in Buildings, Second Edition. Malcolm Holland.
© 2023 John Wiley & Sons Ltd. Published 2023 by John Wiley & Sons Ltd.

Prior to World War One, most domestic floors for living rooms at ground-floor level, were made with timber, with an air space below. In the kitchen, the floor would most commonly have been solid compacted hardcore finished with clay quarry tiles.

During the interwar period it was mostly the same.

During World War Two, a massive cement and concrete industry was developed, to build things like bunkers etc. After the war, there was a shortage of timber. The combination of a shortage of timber and the capacity of the concrete industry, resulted in a change in construction methods. After World War Two, most ground floors became solid concrete.

From the mid-1970s, because of the rising cost of oil and gas, there was financial and political pressure to improve insulation levels in buildings. Subsequent changes in Building Regulations in the United Kingdom, made it more and more difficult to achieve insulation levels using solid concrete-floor construction.

During the 1980s the building industry started to move away from solid floors, to suspended concrete floors, with an air space below. This evolution accelerated after the 1985 revision of the Building Regulations. By 1990 onwards, most houses would have suspended concrete beam and block floors.

It can be quite difficult to tell the difference between a solid floor and a suspended concrete floor. Depending on the floor finish and the spans involved, they can both feel very solid underfoot. However, suspended concrete floors have a void beneath which must be ventilated to prevent the risk of a build-up of methane and radon gas (where applicable). Consequently, a building with suspended concrete floors will have air bricks around the base of the wall. The number of air bricks will vary with the date of the building. Originally, the 1980s requirement was that every void under the floor should be ventilated by a least one air brick on each side of the void. Later changes in Building Regulations increased the number required.

Concrete beam and block floors are made up of upside down 'T' shaped beams, like this, ⏊. The beams span from supporting walls to supporting walls. They are spaced just over 500 mm apart so that the gaps in between are the right size to fit standard length concrete blocks (440 mm long blocks). See Figure 2.29.1.

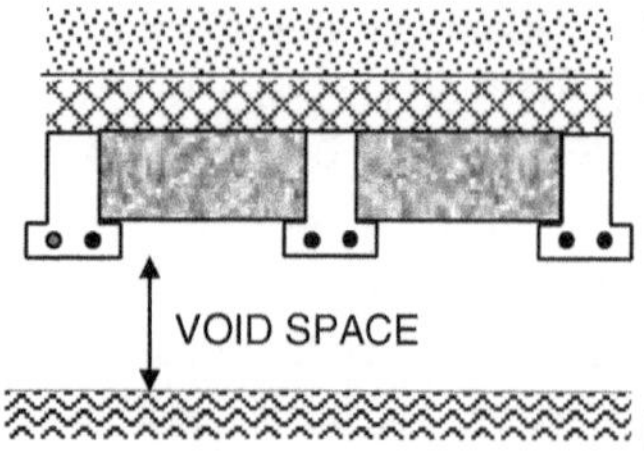

75mm Floor finish cement: sand screed

Insulation board, polystyrene or similar

100mm deep concrete blocks between beams

150mm deep reinforced concrete inverted "T" beams.

VOID SPACE

150mm minimum vented void space beneath floor

Bare earth ground level under floor

Figure 2.29.1 Beam and block floor construction.

Once the beams and blocks have been placed in position, an insulation board layer of polystyrene or similar is laid on top. A surface wearing layer is then laid on top of the insulation. The surface wearing layer can be a flexible material, such as chipboard decking. It is however, more commonly, a rigid cement: sand screed. A screed is a mixture of cement and sand, typically at a ratio of 1:3 or 1:4; mixed with water so that it can be laid and spread. When it is laid on a flexible surface, like insulation board, it has to be quite thick, typically around 75 mm; and also reinforced with a light metal mesh.

The beams are usually 150 mm deep, to be compatible with typical brick courses. Each brick-and-mortar course is nominally 75 mm deep. Two courses are 150 mm. This relatively small depth restricts the possible span length of the beams. The strength of a beam has far more to do with its depth, than its width.

Also, the beams must not be so heavy that they cannot be manually handled. They can be lifted onto the walls by plant machinery, but they must be handled into precise position, straight, square to each other and at the correct spacing, by manual labour.

If the beams are too long, they will become also too 'springy' in the middle.

Consequently, the length of a beam is usually kept to between 3 and 4 m (and smaller where possible).

The footprint of a house is obviously more than 4 m wide in most cases, and therefore several spans are used. The beams span from wall to wall. See Figure 2.29.2.

Where the two spans meet, there is a junction of the beams. Quite commonly, the walls above floor level will be built off the same foundation as the floors and therefore the joint will be covered by the internal wall at ground level. However, where there is a door opening, the floor finish must be carried through from one side of the wall to the other. The floor insulation and floor screed are laid over the junction at the internal door threshold. The screed might also be topped with a rigid decorative finish, for example, floor tiles.

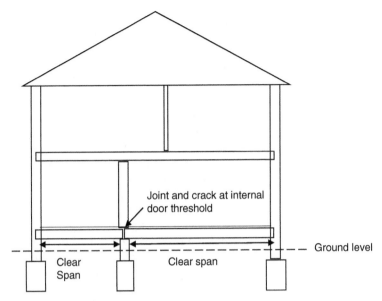

Figure 2.29.2 Typical floor spans.

It is quite common to find a crack across the floor at the threshold of the internal door. This is at the junction of the floor beams beneath. This is caused by one, or a combination, of the following factors.

Firstly, after a building is built, it settles down. The most heavily loaded parts will settle quicker and to a larger magnitude than the lightly loaded parts. This might not be perceptible to human senses but can be enough to fracture rigid finishes such as the floor screed and tiles.

If two parts move at different rates and amounts there can be movement at the junction of the two.

The beams will also expand and contract. Concrete has quite a high co-efficient of expansion. When the two beams expand and contract, they will push and pull at the joint.

The beams are subject to loads as things are placed on them or people move around. The vibration will shake the joints.

If the longer beams deflect under load, more than the shorter spans, it will lever the joint open.

Any one of these movements, or a combination of them, causes the floor to crack.

The cracks will usually be quite small, hairline to a millimetre or so. There can also be a slight change in level or 'lip' at the crack. Although the movement and cracks are small, it often has a disproportionate aesthetic affect, for example, by breaking the floor tiles in a kitchen.

If the cracking is repaired, it is likely to recur due to cyclical movements such as temperature and other seasonal variations. Although the cracks will recur, the magnitude will not become greater with time and there is no risk to the structure of the building.

One way to avoid recurring cracking is to put a movement joint in the floor finish. Fitting a 10 mm wide metal strip across the threshold will provide a movement joint that will stop floor tiles from cracking. Metal strips come in various colours, specifically for this purpose.

Less commonly cracking may also occur at external door thresholds. The beams are supported on the inner leaf of the wall. At external door thresholds there is a gap between the end of the beam and the position of the door frame. This gap, over the wall cavity, has to be infilled on site by cut bricks or concrete. The screed topping is then carried over the infilled gap to the back of the door frame. Cracking sometimes occurs in line with the junction of the beam and block floor and the infilled part because of differential movement, as described above.

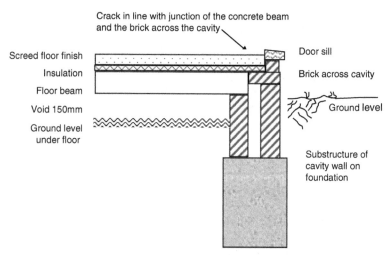

Figure 2.29.3 Crack at external door threshold.

2.30

Vehicular Impact Damage (Usually Garages)

Vehicular impact damage can occur to buildings, particularly garages. Surveyors should remain alert to this possibility in vehicular trafficked areas. The most at risk areas are the end wall, piers by doors and brickwork above the door. Cracking can be confused with subsidence. The severity depends on the force and frequency of impact. Damage is most commonly minor but can be severe. Severe damage may require partial rebuilding.

Key features

1. Rear walls pushed out at car bonnet/bumper height.
2. Rotational movement causing stepped cracking.
3. Vertical cracking adjacent to door piers.
4. Gable walls and lintels over entrance doors displaced.
5. Localized, direct damage at impact points.

When inspecting movement and cracking in buildings, it is often easy to overlook the obvious. In areas where vehicles are used, the obvious cause of movement is often collision damage.

Garage entrance doors are often quite small in relation to the size of cars and vans, leaving little room for misjudgement. Garages are rarely

Practical Guide to Diagnosing Structural Movement in Buildings, Second Edition.
Malcolm Holland.
© 2023 John Wiley & Sons Ltd. Published 2023 by John Wiley & Sons Ltd.

much longer than a car or van, and it is easy to not stop quite in time and bump the end wall.

So, when inspecting a garage (or any building exposed to vehicular traffic), it is a very real possibility that a vehicle has simply impacted it. The impact could be almost anywhere, but the most common collisions involve hitting the end wall or the brickwork around the door.

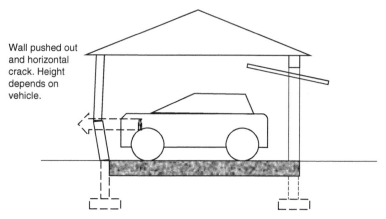

Figure 2.30.1   End wall collision.

If a car drives into a garage and hits the end wall, it can push it outwards. The magnitude of the movement will depend on the force of the impact and the frequency that it occurs. It may of course have been impacted several times over the life of the building. Typically, a horizontal crack will occur at the line of the impact. A relatively small impact will tend to bow the brickwork outwards. A significant impact will displace it outwards at the impact line, causing a step out in the brickwork.

The height of the impact will of course depend on the height of the vehicle. People tend to change cars every few years, so the wall may have been hit by different cars, at different heights, over several decades.

In most cases the impact will be light and infrequent.

The effect of pushing the end wall out can also cause stepped cracking in the side wall which could easily be confused with subsidence. See Figure 2.30.2.

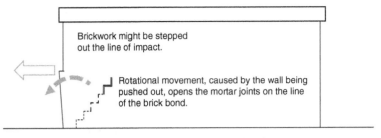

Figure 2.30.2 Stepped cracking caused by rotational movement.

As the wall is pushed at the point of impact, it overturns the wall to some degree. This rotation pulls the brickwork open along the mortar joints following the bond of the brickwork. The crack will be slightly wider at the top than the bottom because the wall will have been pushed out most at the impact level. This will give a diagonal crack that is wider at the top than the bottom, much like a typical subsidence crack. If some stepped cracking is seen in this position on a garage, check the end wall with a spirit level to see if it has been pushed out of vertical by impact. If the cracking is just related to this position and the end wall has been pushed out, impact is the most likely cause, not subsidence (unless of course there is some other reason to suspect subsidence risk, for example trees nearby on shrinkable soil).

This type of damage is not progressive, unless of course it is driven into again!

In most cases repair will involve simple repointing of the brickwork. If the end wall has been significantly knocked out of vertical it might become unstable. In which case it would need to be taken down and rebuilt.

If the impact is severe, the vehicle may crash right through the wall. In which case it would need to be re-built.

The other area that often suffers most from impact damage is the jambs to the doors. Garage walls are often built with single skin brick walls, 100 mm thick. The walls are thickened at the door openings to form piers. The piers provide lateral stability to the walls and extra strength to support the doors. Mid-length along the walls, there are also usually stiffening piers of brickwork to provide lateral stability to the thin brickwork. See Figure 2.30.3.

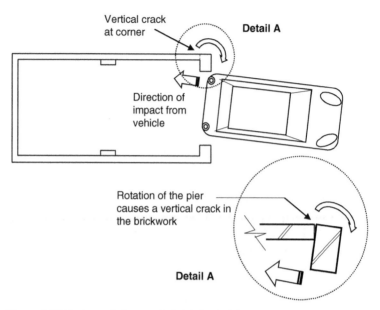

Figure 2.30.3   Impact damage to brick jambs/piers.

It is not uncommon for vehicles to hit one of the brick piers at the entrance to the door (see Figure 2.30.3). This can cause direct damage to the brickwork, which can cause local random cracking or shattering of the bricks and mortar. A common crack pattern that often occurs, is a vertical crack in the brickwork and mortar, where the pier of the brick-work joins to the single skin of the rest of the wall. The cracking will often be full height but probably be widest at the point of impact. The point of impact depends on the height of the vehicle. As a vehicle hits the corner of the pier, it rotates it. It is this rotation that causes the crack. The width of the crack will depend on how hard it was hit and how often it has been hit.

If this damage is minor, it can be re-pointed. However, it is quite common for the brick pier and the wall to become separated and loose. The brick pier in the wall, is there to provide lateral stability to the corner. If it is loosened, it no longer provides this. If this occurs, the pier should be taken down and rebuilt, bonding it back into the brickwork of the wall so that is returned to being an integral brick pier.

A less common form of impact is where a high vehicle, too high for the door opening, hits the lintel and gable wall above the door. See Figure 2.30.4.

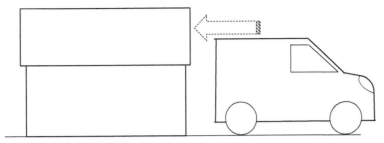

Figure 2.30.4 High vehicle hitting the lintel and gable above the door.

This would typically be a high van or lorry.

The lintel or brickwork (or both) would be pushed in causing localized damage. The repair work required would depend on the severity of the impact. Minor localized impact damage could be repointed. If the lintel is knocked partially off its bearing, the gable wall would need to be taken down so that the lintel could be re-instated. The gable wall would then be rebuilt.

In older garages built pre-1970 usually, the roof might have timber purlins (beams) built into the gable wall to support the roof. In which case, temporary support to the roof would be required during remedial work. One advantage of having the purlins built into the gable, is that it holds the gable in place. This would provide some resistance to impact damage.

In most modern garages, post-1970 approximately, the roof is likely to be modern truss rafters. With this type of roof there is no load transferred to the gable wall. The gable wall is therefore unloaded, other than its own self-weight. It is therefore easily moved if it is impacted.

If the brickwork is knocked out of vertical it could become unstable and again this would require re-building works to be undertaken.

The extent of damage and repairs would depend on the facts in each case.

Part 3

Cracks in Buildings Related to the Foundations and Ground Movement

3.1

Introduction

The purpose and functional requirements of foundations are described. A brief history is included to describe how foundations have developed over the years and why they have become deeper and more substantial with time. The difference between the terms 'settlement' and 'subsidence', is explained.

Foundations have to be able to support the weight of a building, plus any imposed loads such as wind and snow. They must be able to support these loads without the material of the foundation breaking. The foundations then transmit (distribute) the weight to the ground.

Foundations are usually set below ground. This is to avoid seasonal movement from frost and soil moisture content changes. If water in soil freezes, ice will form, and this ice can cause an expansion in volume. The expansion might lift the foundations. In some clay soils, changes in moisture content can cause it to swell up when wet, and shrink when dry. These changes occur more frequently, and to a far greater degree, near the surface than they do at depth. As a general rule, the deeper the foundations, the better they are likely to be.

Foundations set into the ground are exposed to chemicals within the soil. The material must be able to resist chemical attack. They are also likely to be exposed to the effects of frost. The material must also be durable to resist this.

Practical Guide to Diagnosing Structural Movement in Buildings, Second Edition.
Malcolm Holland.
© 2023 John Wiley & Sons Ltd. Published 2023 by John Wiley & Sons Ltd.

Before the advent of mechanical digging equipment, it was a hard and laborious job to dig deep foundation trenches. As a result, foundations in most houses built before World War One were most commonly no more than about 300 mm (1 foot) deep. The exception is houses with cellars where the foundations are below the cellar walls and therefore, more than 2 metres beneath the external ground level.

Interwar houses, between 1918 and 1939, will probably have foundations between 450 mm–600 mm deep (1½ to 2 feet).

After World War Two, the standard is about 900 mm (3 feet, 1 yard). Nowadays this is usually interpreted as one metre.

One has to ask why we go to the expense of building foundations one metre deep, when the vast majority of historic housing has performed so well, over decades or centuries, with much shallower foundations. This is largely explained by a change in how movement in buildings is perceived. Historically some movement was simply tolerated, as long as it did not develop and become dangerous. Even today, historic movement in old cottages is often considered to be part of their charm. A significant proportion of the population prefer older houses for this very reason. The same view would not be taken of the same degree of movement, in a modern building.

As a general rule, it would arguably be better for the economy of the country as a whole to build much cheaper foundations. More substantial foundations would only be needed to suit specific site conditions when these were poor. Even if less substantial foundations led to an increase in movement, it would not endanger the performance of the building in most cases.

The increase in the number of serious failures would be a very small percentage and these could be repaired. The cost of the repairs could be borne by insurance, for which we would all pay in aggregate, by increased policy fees. That way the net aggregate cost on the economy would be lower and no single individual would suffer catastrophically. This would, however, require a major change in attitudes.

The point of this discussion, however, is just to emphasize that foundation design is not set in stone, or more commonly nowadays concrete. Foundation design is not just a matter of the technical solutions either. There is some emotional investment, in the way in which foundations have evolved over the years.

It is worth mentioning here the difference between the terms 'SUBSIDENCE' and 'SETTLEMENT' in relation to foundations.

Subsidence is movement of the foundations related to the ground. That means it is not caused by the weight of the building. For example, when clay soil shrinks, causing the foundations of a building to subside, this would happen whether it was a bungalow with little load or a much taller building with a lot of load. The movement is caused by the ground moving beneath the foundations.

Settlement is movement caused by the weight of the building. When a new building is built, the weight of the building compresses the soil beneath it. The compression of the soil forces out moisture and air voids. This makes the soil stronger, until the strength of the soil matches the load applied on it. Most of this movement takes place as the building is being built or soon after it has been built. Some long-term creep movement will, however, occur after construction, particularly in clay soils.

If, for example, part of a building is two storeys high and another part is only one storey high, the weight on the soil will be different under the two parts. The ground under the two storey section will be compressed more than the ground under the single storey section.

Sometimes minor cracking will occur at the joint between the two as a result. An old building, which has had time to settle down, will have reached equilibrium. If an extension is added, it will be some years before that extension reaches its equilibrium. As a result, there is likely to be some differential settlement between the two.

When a new opening is formed in a wall, the weight is transferred around the opening via a beam. This weight is then concentrated at the support points. This load concentration will cause some movement in the property, by compressing the soil at the support points. The load causes the soil to become more compact and makes it stronger. After a short time, hopefully before significant movement occurs, a new equilibrium is reached.

3.1.1 Design For Load

The purpose of foundations is primarily to distribute the weight of the building to the soil. Historically, the assessment of soil strength was crude and often just a guesstimate, based on visual inspection and experience. There are more sophisticated soil tests available today but soil is a variable material. Even with the test methods available, a high design margin of safety has to be allowed for.

The foundations of a building serve the primary purpose of transferring the weight of the building to the subsoil. In order to do this, they have to be large enough to spread the considerable weight of a building, over a large enough area of subsoil.

The size of the foundations is therefore determined by the strength of the soil and the weight applied to it.

There are a number of standard tests for soil to determine its strength. These tests are not particularly sophisticated. For example, imagine dropping a metal spike into the ground. Provided the spike was always the same weight and was always dropped from the same height, the depth to which the spike penetrated the ground would depend on the density of the soil. There is a relationship between the density and the strength of the soil. This is the basis of the standard penetrometer test. A British Standard spike, dropped from a British Standard height, into the ground.

There are other tests that can be carried out on site or based on disturbed samples.

Soil is not a manufactured uniform material. It is not difficult to imagine that results from one part of a site may vary from another part. Also, several tests carried out, on what would appear to be pretty much the same soil, would give a spread of results. In order to allow for this variance, a large design factor of safety is applied in the design process. The result is a design margin of somewhere around three. That is, the foundations are designed only to accept a load about one third of what the soil could possibly be expected to take.

Historically no tests other than a visual inspection would take place. The soil would be visually inspected and described in broad terms, such as, how hard it is to dig with a shovel and whether it is sandy or clay. From this rough assessment, a range of the likely bearing capacity can be assumed, based on recorded experience from the past. Provided that the assumed bearing capacity is kept within the bottom part of the range and additionally a factor of safety is applied on top of this, experience has shown that it is reasonably reliable.

The most common subsoils, clay and sand, are generally speaking, easily strong enough to support brick-built buildings up to three storeys high without any sophisticated or specialist foundations. In parts of the country where there are weaker or less reliable soils, foundation design may have developed empirically to overcome this. Timber piled foundations in wet ground are perhaps one example of this.

3.1.2   Design For Stability

> *Foundations are designed to be unaffected by changes in the subsoil. This is usually achieved by being set at a depth where frost heave or changes in moisture content will not affect them. The depth of a foundation is more likely to be determined by stability considerations, than it is load bearing capacity.*

Experience over time has shown that foundations do not necessarily have to be very deep below ground to be able to carry the load of the building. Depth is, however, a very relevant factor when considering stability. It is the desire for stability that largely accounts for the design depth of foundations.

The moisture content of the subsoil will vary seasonally. The amount of water in the subsoil will affect its volume. Clay soil will expand when wet. It will shrink when dry. The amount of swelling or shrinkage depends on the particle size of the clay and the 'purity' of the clay. Subsoils are often a mixture of different types of soil. For example, a predominantly clay soil, may contain a percentage of sand, gravel or stone mixed in. Where there is a mix of clay and solid material, such as stone, only the clay element of the mix will change volume. A soil that is 100% clay will be more prone to shrinkage than a part clay soil, as the whole volume will be subject to moisture content changes.

The particle size of the clay can also vary. The smaller the particle size the greater the propensity to volume changes. This is known as the 'Plasticity Index'. Clays are banded depending on the Plasticity Index as high, medium or low shrink-ability.

Experience has shown, that under normal conditions, with the absence of any significant trees or large shrubbery, clay soils are not usually significantly affected below a depth of one metre. That is why a 'standard' foundation depth of one metre in a clay soil, is considered to be acceptable for Building Regulation purposes. Even in clay soils, foundations would only need to be deeper than this if trees were present.

In sandy or gravel soils, the volume is not greatly affected by seasonal water variations. There is, however, the risk of frost heave if moisture in the soil freezes. A depth of 450 mm is usually sufficient to prevent the risk of frost heave. In such subsoil conditions, a depth of one metre is not essential. A depth of 450 mm would be adequate, provided of course that the soil at this depth had the necessary bearing capacity.

If the subsoil is bedrock, this would not be affected by frost or moisture content variations. Once bedrock has been reached this would be the acceptable depth for stability and load.

Contrary to what most people believe, the Building Regulations do not state that a foundation should be set at a specific depth. The Regulations are written more intelligently than this. They simply require that foundations should be set at an adequate depth to avoid seasonable ground movement and to have sufficient strength. They are a performance requirement not a prescription.

This gives the designer the freedom to design foundations in whatever way is seen fit, provided of course that they can perform the function demanded of them.

3.1.3   Identifying Below Ground Defects

Simple visual inspection may not always be enough to diagnose an underground defect. It is impossible to see below ground. Historical events may have occurred which are lost in the mist of time. Whilst in many cases the answer will seem obvious there will always be some cases where the initial visual inspection can only narrow down the likely causes.

Identifying and diagnosing defects in buildings, where the cause is below ground, has an added difficulty over above-ground defects. That is, it is impossible to see underground.

It is often a matter of narrowing down the possibilities and arriving at the most likely explanation during the initial visual inspection. Further investigation may then be required, in order to confirm the initial opinion (or not as the case might be).

For example, if there is a crack in a building that looks like it might be related to subsidence then the question is why?

If the ground is clay, there is a large oak tree nearby, it is the end of a drought summer, the building is old (and therefore the foundations are likely to be shallow) and the imaginary arrows of tension point to the tree; then it is reasonable to draw the conclusion that clay shrinkage is the most likely cause.

If the ground is sandy, there is isolated movement at a corner, and at the same corner there is a downpipe discharging water onto the ground, or into below-ground drainage that is 200 years old; then it would be reasonable to form the opinion that this drainage is the most likely cause.

If there is no visible reason above ground at all but an area is known to be subject to mining subsidence, and other properties in the street show similar signs; then it might be reasonable to suspect this as the cause.

In any one of these examples, the opinion may turn out not to be correct. The foundations could have been built over a section of historically made-up ground that has compressed locally. Alternatively, the foundation may have been built with substandard materials that have deteriorated. The walls may have been built off centre of the foundations and they could have rotated. Any one of these three possibilities could give similar symptoms visible above ground.

Without X-Ray eyes it is impossible to know what is hidden beneath the ground. In most cases it is impossible to know what historical events may have occurred which have been lost in the mists of time.

Just because an opinion turns out to be wrong, it does not necessarily make the opinion negligent. Provided it can be demonstrated that a reasonable process has been followed, with reasonable care and skill, an opinion that turns out to be wrong is not negligent. These things often come down to the balance of probabilities. The defect might have two or more possible causes. If an initial investigation can only narrow it down to one or two possible causes, investigate further to confirm the most likely and/or to eliminate the possibility of the less likely.

The cost of repairing foundations is usually high and such costs are best avoided, unless they are really necessary. The initial opinion must always be confirmed before committing to works.

Further investigation often involves excavation. For example, if the initial opinion is that it is probably a drain gulley next to the wall, excavation may need to be carried out around the gulley to check it. A trial pit might be needed to check the depth and construction of footings.

An initial visual inspection will rely on local knowledge and geological survey maps to determine the likely nature of the soil. It might then be necessary to dig trial pits or take samples using an auger, to determine the nature of the subsoil more accurately and specifically to the location.

If a defect looks serious, in magnitude or speed of progression, and if the movement is considered to be related to below-ground movement, the visual inspection and opinion forms only the first part of the identification process.

Before committing to remedial work that might be expensive, the movement may need to be monitored for some time and further investigation work will probably be necessary.

Foundation settlement and subsidence cracks often share some common features.

Key features

1. 45° crack through the bond of the bricks.
2. Cracks through the junction of the mortar and the bricks are most common but if the mortar is strong the crack can go through the bricks.
3. Cracks externally and internally.
4. Cracks get wider at the top because of a rotational effect.
5. The crack will continue beneath the damp-proof course and go below ground.
6. Distortion of door and window openings.
7. Likelihood of external factors, trees, drains, made ground, mines, etc.
8. Movement can be small or large, hairline or 5 mm, 10 mm, 20 mm or even collapse.

Foundation Movement Caused by Clay Shrinkage

Much of the subsoil in Britain is clay, particularly in the Southeast of England. Clay soils undergo changes in volume in relation to their moisture content. Trees extract moisture from the soil. During extended periods of dry weather, the water is not replenished. This can lead to clay shrinkage and subsidence in buildings. The risk of movement depends on how shrinkable the subsoil is, how close the tree is, the tree species and the depth of the foundations. Some trees represent a particular risk. These are Oak, Willow, Poplar, Eucalyptus and Elm. Any large tree is a risk. Fruit trees tend to use a lot of water during the fruiting season. Whilst all cracks will start from zero, cracks related with clay shrinkage subsidence can quickly become large, 5 mm, 10 mm or even wider. The crack pattern is a generally diagonal at about 45° from vertical. Cracks are tapered, becoming wider with height. Displacement across the crack is similar both horizontally and vertically. Movement in a building will recover to some degree when the soil becomes wet again. Remedial work may involve reducing or removing the tree/trees, together with just cosmetic repair. In more severe cases it can require underpinning of the foundations or even demolition.

Practical Guide to Diagnosing Structural Movement in Buildings, Second Edition.
Malcolm Holland.
© 2023 John Wiley & Sons Ltd. Published 2023 by John Wiley & Sons Ltd.

Key features

1. Diagonal cracks at 45° (subject to opening positions).
2. Tapered cracks become larger with height.
3. Crack displacement both horizontally and vertically.
4. Cracks extend below damp-proof course and ground level.
5. Cracks start from zero but commonly quickly enlarge to 5 mm, 10 mm or greater.
6. Cracks internally and externally.
7. Clay soil.
8. Presence of trees, bushes or hedges.
9. Increase risk with age of buildings as foundations are more likely to be shallow.
10. Dry weather, usually drought.

Much of the subsoil in Britain, particularly in the South of England, is clay. Where the subsoil is clay, the most common form of foundation movement is subsidence due to clay shrinkage. Clay shrinkage is usually a result of moisture extraction caused by trees and other plants.

The level of risk is determined by:

1. How shrinkable the soil is.
2. The depth of the foundations.
3. The type, size and species of tree.
4. The severity of any drought weather.

As has already been mentioned, the amount by which the soil will shrink or expand depends on the particle size of the clay and the percentage of clay in the soil. A soil that is entirely clay will shrink more than a soil that has only 50% shrinkable clay material within it. The smaller the particle size of the clay, the greater the volume will change with moisture content.

The soil will dry out from the top down and therefore the soil shrinks most at the surface. It is common to see very large cracks in the surface of clay soil during the summer. These quickly close up when wet weather returns. The greater the depth of the foundations, the less likely it is that the moisture content will change to a significant degree.

Without trees nearby, modern one-metre deep foundations are unlikely to be affected, even in drought conditions. Even much shallower foundations are rarely affected. If they do go up and down a bit seasonally, this usually has no perceivable effect, provided that the movement is uniform across the whole building.

In order for volume changes to be significant at foundation depth, there usually has to be a tree, hedge or bushes.

Some trees are classed as high moisture extractors. The most significant are Oak, Willow, Poplar, Elm and Eucalyptus. As a rule of thumb, the larger a tree is and the greater its leaf crop, the more moisture it will extract. Any large tree is therefore a potential problem. Big trees such as Chestnut, Sycamore, and Beech, amongst others, may be a greater threat than smaller trees that have a high moisture extraction classification, purely because of their size. Fruit trees generally also require a lot of water to fill the fruit and are also quite commonly associated with this type of movement. Conifers with needles rather than leaves are low in water demand. Conifers like cypresses with feathery leaves are relatively medium demand. Some species of conifers will quickly grow to a large size, the most well known being Cypress Leylandii.

Even allowing for this, most modern foundations of one metre depth are able to weather a single dry or drought summer without movement. It is usually only when there is a period of dry summer, dry winter and then a second dry summer, that the number of problems increases significantly.

A traditional rule of thumb was that the height of a tree should not exceed its distance from the property. This applies to medium water demand trees. With a high moisture extraction species this could be extended to the distance being 1.5 times the height of the tree. Where there is a row of trees, again the rule of thumb would be that the distance should be 1.5 times the height of the trees. With trees having a low water demand this could be reduced to 0.75 times the height of the tree. Tables are published by a number of sources which give a more accurate assessment, by tree species, but it should be recognised that this is not a precise science.

This rule of thumb is reasonably effective and reliable where the foundations are a metre deep.

To be more accurate than this, it would be necessary to carry out a more detailed analysis of the soil to determine its plasticity index. The tree species would also need to be determined. With the results of soil analysis and identification of the tree species, it is possible to design the depth of foundations to mitigate the risk.

When clay shrinkage occurs at depth, the foundations will subside. This will cause a downward movement and also a rotational movement. The rotational movement is created by the soil shrinking on one side (the tree side) more than the other side. The wall will sink and rotate until it 'hinges' and cracks. The cracks will roughly be at about 45°. This angle may be distorted by the position of openings.

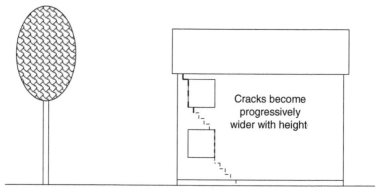

Cracks become progressively wider with height

Crack extends below damp proof course and ground level

Figure 3.2.1   Clay shrinkage crack pattern.

The displacement of the cracks will be both vertical and horizontal (Figure 3.2.1). The displacement should be roughly equal horizontally and vertically in an idealized example, but in reality, gravity will tend to close up the vertical displacement. Rotational movement will also exacerbate the horizontal component of the movement. It is, therefore, normal to find the horizontal width of the crack is greater than the downward gap. This horizontal displacement increases with height due to rotation. Therefore, the crack will be tapered, increasing in width with height.

The cracks are likely to pass through and below the damp proof course and then disappear below ground level. The crack at the base of the wall may, however, not be visible. The mortar below damp proof course level is damp and therefore, a bit softer and more flexible. The crack may also become very narrow, to the extent that the damp mortar can accommodate the movement and the crack becomes invisible.

A modern plastic or bitumen strip damp proof course can also act as a slip membrane. The crack can be transferred from its expected 45° route, several metres along the wall, often to a door opening position, should one be present.

Where services enter or leave a building, there has to be an opening in the wall or foundations below ground. Underground openings may also create a 'hinge' point, to where the crack below damp-proof course level may be transferred, even though there is no visible opening like a door. The position of an underground drain passing through the foundations will usually correspond with the position of sanitary fittings or waste pipes internally.

Obviously, the crack width will start off from nothing and progressively enlarge as the soil shrinks more and more during the drought period. Cracks can, quite quickly, progress to 5 mm, 10 mm, 20 mm or even greater, depending on the variables of soil type, tree species and weather. If the drought weather ends before the crack has progressed, it may just stop at hairline.

In severe cases, stability can become affected, requiring temporary buttressing with large raking timber/steel shores to prevent collapse.

The general soil type can be determined by reference to a Geological Survey map or online data. The knowledge of the local Building Control Department, who regularly inspect foundation trenches, is also invaluable. Precise knowledge can, however, only be obtained by taking a soil sample for laboratory analysis, to determine the nature of the soil and how shrinkable it is. This sample can be taken by auger or excavation.

The tree species can be identified by reference to one of the many textbooks, websites or phone apps, available on tree identification. By examining the leaves, twigs, bark, overall size and shape; and comparing this to the details in the books, websites or phone apps, the tree type can be determined. Alternatively, if the tree species is not obvious consult an arborist.

The repair method depends on the severity of the damage. If movement is identified quickly, the offending tree is relatively small, and not protected by legislation, it can be removed. The soil should then recover. There can, however, be risks associated with the removal of trees. Clay can swell up causing heave damage to buildings. Please see also Chapter 3.3 Clay Heave.

After trees have been removed, the cracking can be monitored until it has stabilized. The cracked area can then be re-pointed. Sometimes an epoxy mortar is used to re-point the wall in order to achieve a higher tensile strength and prevent the risk of cracking recurring on the same line. With large cracks, over 5 mm, sometimes the horizontal mortar joints are raked out and horizontal metal reinforcing is also inserted into the joints to give increased tensile strength before re-pointing.

If it is not desirable to completely remove a tree, it could be cut back in order to remove its leaf mass and overall capacity to transpire moisture.

If the tree or trees are protected by legislation, for example, a tree preservation order or by being within a conservation area, it will not be possible to remove it/them or even to cut them back significantly.

Sometimes a concrete root barrier is built. This involves digging a trench between the building and the tree/trees. The trench is then filled with concrete to form an underground barrier.

If the movement is severe, or if the tree has to remain in place, then under pinning may be required. This involves building new, deeper foundations under the existing building. There are a number of ways that this can be achieved, which are briefly described in Part 4 'Repair Methods'.

Whatever method is employed, it is a disruptive and expensive operation that is best avoided if cheaper alternatives are available. In very severe cases, underpinning and partial rebuilding may be required. Demolition may also be an option in severe cases.

If movement occurs in a building, late in the dry season, and does not develop to a serious condition, it will commonly just recover naturally when the weather turns wet again.

The problem of clay shrinkage is related to how much water is discharged into the soil, as well as how much is extracted. Historically, or even relatively recently in non- urban areas, a significant amount of rainwater would simply have been discharged onto the ground. Rainwater was often collected from roofs and stored in tanks or butts to water garden plants. Underground soakaways were also common.

Over time, much of the water that would have entered the ground has been removed. As urban areas expanded rapidly during the industrial revolution period, rainwater from large urban developments was piped away in combination with foul drainage. After World War Two, it became common to build two drainage systems, one for the surface water disposal and one for foul waste. Hard landscaping for roads and pavements has also significantly reduced the amount of rainwater that can enter the ground. Most of the rainwater that falls, particularly in urban areas, runs off very quickly and is piped away. The net effect of this is less water in the ground for trees to access. There is also an increased risk of flooding, as surface water drainage pipes quickly send a heavy surge of water to rivers and balancing lakes.

It was not until around the mid 1990s that this problem started to be addressed and the term 'sustainable urban drainage' or 'sustainable surface water drainage' came into common use. This involves building an underground soakaway or several soakaways to collect rainwater from roofs and hard surfaces. The chambers are often quite large. To prevent them from collapsing, they are usually filled with hollow plastic lattice boxes, which look similar to milk crates. The underground soakaways have an overflow to the drainage system for periods of very heavy rain, but during light or normal periods the rainwater is released naturally by soaking into the ground.

Where buildings are within influencing distance of trees, it might be sensible to collect rainwater from roofs and surfaces and deliberately pipe it to the risk areas. The bowl of ground under the canopy of the tree is very dry, as the tree shelters it from rainfall. Over wetting the soil in dry risk areas could create a 'buffer' against drought conditions. Even during dry summers, it is not uncommon to get very heavy rain during thunderstorms. If this water could be harvested, rather than just run off into a drain, it might just provide enough water to prevent damage or even to cause recovery.

A short-term treatment, just to get through the season, might be to water the ground. Unfortunately during a drought period, a hosepipe ban will probably be in effect preventing this in most cases.

Figure 3.2.2 Severe well-developed diagonal cracking caused by clay shrinkage subsidence and the proximity of trees.

Figure 3.2.3   Close up of diagonal subsidence cracking. Displacement both horizontal and vertical.

3.3

Clay Heave

Much of the subsoil in Britain is clay, particularly in the Southeast of England. Clay soils undergo changes in volume in relation to their moisture content. Trees extract moisture from the soil. When a tree or trees are removed, the soil will take up moisture and the clay will swell. Swelling can take place for some years after removal of the trees, until a new equilibrium is achieved. The swelling of the soil can lift buildings, particularly ground bearing floors. This is often associated with new construction where trees and hedges have been removed from the site during site preparation. It can; however, affect any age of building, if a tree is removed. The magnitude can vary widely from just cosmetic damage to severe structural damage.

Key features

1. Diagonal cracks at 45° (subject to opening positions).
2. Crack displacement both horizontally and vertically.
3. Cracks extend below damp-proof course and ground level.
4. Cracks start from zero width, but 5–10 mm not uncommon.
5. Cracks internally and externally.
6. Clay soil.
7. Often new build or new extensions.
8. Following removal of trees, bushes or dense shrubs.

Practical Guide to Diagnosing Structural Movement in Buildings, Second Edition.
Malcolm Holland.
© 2023 John Wiley & Sons Ltd. Published 2023 by John Wiley & Sons Ltd.

Just as clay will shrink when moisture is removed from it, it will also expand when moisture is put into it. This most commonly occurs when trees are removed from a site prior to construction of a building. Over a few years, following the removal of trees, the subsoil will absorb moisture and expand until a new equilibrium level is reached. Similarly, if a tree, hedge or substantial shrubs are removed from the proximity of an existing building, the change in moisture content can lead to heave damage. The swelling will push up the building, particularly ground bearing floors which are not as heavy as the walls.

The appearance of the cracks will often look similar to foundation movement cracks caused by shrinkage of clay, but this does depend on where the soil expands in relation to the position of the building. For example, in Figure 3.3.1, trees have been removed to the right of the building.

As a result, the soil has expanded on that side lifting the right-hand section of the building.

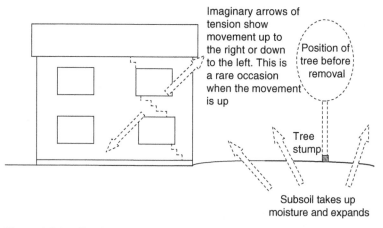

Figure 3.3.1   Clay heave at side of building.

Imaginary lines of tension at right angles to the crack, point down to the left or up to the right.

In this example there is no obvious reason why the building should have moved down to the left. The presence of a tree stump is a reason why it might have moved up to the right. In the absence of any other evidence, it would be reasonable to assume this is the most likely cause.

In practice when trees are removed, the stump is usually left in place and poisoned to prevent it from re-growing. This is cheaper and less disruptive than digging it out. Over years and decades, it will rot away.

In Figure 3.3.2, imagine a situation where the heave is more directly under the building.

As this type of movement is related to the removal of trees, hedges, bushes or dense shrubs, it is often related to new construction or the construction of extensions. In such cases it can cause substantial damage for several years after the time of construction. The rate of movement will be a relatively rapid expansion initially, after which the rate of expansion will slow down over time, until a new equilibrium is reached. It may take up to around five years to reach a reasonable equilibrium. Provided that the magnitude of movement is not great enough to require demolition or major re-building, the cracks can be simply re-pointed or cut out and stitched back together with an epoxy mortar.

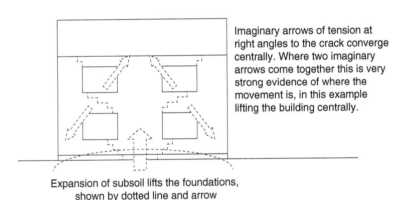

Imaginary arrows of tension at right angles to the crack converge centrally. Where two imaginary arrows come together this is very strong evidence of where the movement is, in this example lifting the building centrally.

Expansion of subsoil lifts the foundations, shown by dotted line and arrow

Figure 3.3.2 Clay heave under building.

Heave movement may also occur when trees are removed from nearby existing buildings. Sometimes the trees are removed following the first signs of clay shrinkage. See Chapter 3.2 Foundation Movement Caused by Clay Shrinkage.

Clay heave cracking is often found in concrete garage floors. Concrete garage floors are ground bearing. Ground bearing means that they are cast as a slab of concrete usually 100mm to 150mm thick on a 150mm thick bed of hardcore. The depth of the floor and the hardcore together is therefore 250-300mm. The weight on the floor is much less than the weight of the walls on the foundations. Any swelling of the ground can more easily lift the floor than the walls. So even where there is only a little shallow level heave it can lift the garage floor slab causing it to crack. The cracks are usually minor and have little or no effect on the practical use of the floor as a garage.

3.4

Seasonal Expansion of Clay Subsoil

Clay subsoil will shrink and expand in volume due to changes in the soil moisture content. Even where this is within normal limits, it will cause the building to lift slightly during wet weather and sink slightly when it is dry. The lightly loaded parts of a building can be lifted more easily than the heavily loaded parts. The more heavily loaded parts of a building will sink to a greater degree than the lightly loaded parts. This differential movement causes minor cracks particularly under ground floor windows and patio doors. The cracks are cosmetic in nature. They are non progressive but if re-pointed they will recur seasonally.

One of the most common causes of minor cracks in buildings can be attributed to normal seasonal variations in moisture content within the soil. The building is moved up and down seasonally as the moisture content of the soil changes. Where the loads are evenly distributed, along the foundations, this does not normally result in cracking. Where there are big window openings at ground floor level, or patio/French door openings, it is not uncommon to find small, mostly vertical cracks centrally beneath the openings (Figure 3.4.1).

The weight of the building is transferred around the openings in the wall and passes down the jambs at each side. If the load cannot spread out

Practical Guide to Diagnosing Structural Movement in Buildings, Second Edition. Malcolm Holland.
© 2023 John Wiley & Sons Ltd. Published 2023 by John Wiley & Sons Ltd.

again, through the corbelling of the brickwork, before it reaches foundation depth, there will be little or no load under the opening. Seasonal movement can lift this lightly loaded area more easily than the heavily loaded area. Cracks caused by such normal seasonal movement are usually hairline to 1 mm. They are not progressive but, if re-pointed, they will recur seasonally.

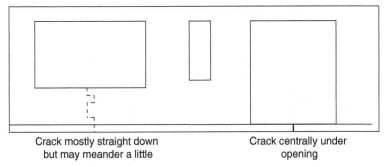

Crack mostly straight down   Crack centrally under
but may meander a little   opening

Figure 3.4.1 Seasonal expansion of clay soil under a window or door opening.

3.5

Eccentric Loading on Foundations

Traditionally foundations were dug by hand. Bricklayers built the walls from within the trenches, corbelling out the bricks at the base to make the footings. After World War Two, foundations were excavated mechanically. Footings were created by pouring concrete into the trenches. This developed to filling the whole trench with concrete, to form trench-fill foundations. This separated the trade constructing the foundations, from the trade building the walls. If the foundations are not set out correctly, the walls may not be built on the centre line of the foundations. This causes an eccentric load. If this is of sufficient degree, it can tip over the foundations. The movement is very rotational, giving a diagonal crack probably below 45°. The horizontal component of the cracking is also wider than the vertical component.

Key features

1. Diagonal cracking.
2. Internal and external cracks corresponding.
3. Much wider cracks at high level than at the base of the wall.
4. Rotational effect can reduce the overall crack angle below 45°.
5. Local movement in one part of the walls.
6. Generally, post-World War Two properties with concrete footings.

Practical Guide to Diagnosing Structural Movement in Buildings, Second Edition.
Malcolm Holland.
© 2023 John Wiley & Sons Ltd. Published 2023 by John Wiley & Sons Ltd.

Traditionally, foundation trenches were dug by hand. Bricklayers then built the walls in the trenches from ground up.

After World War Two, there was a shortage of skilled labour and a large post-war concrete industry, which no longer had fortresses to build. As a result, it became normal practice to build foundations by digging a trench in the ground with an excavator, and then filling the bottom of the trench with concrete. This is known as a strip foundation. Over the years this has developed, and it is now normal practice to fill the whole trench with concrete to save on labour time. This is known as a 'trench fill foundation'.

The construction of the foundations and the construction of the walls, by a single trade operative, have become separated by this work practice.

The walls should be positioned centrally on the foundations for stability. If, however, the bricklayer finds that the foundations have not been built in exactly the right place, not exactly square or not in a straight line, it may not be possible for this to be achieved.

It would be difficult and expensive to alter the foundation after they have been cast. In many cases it would be impractical in any case. It would be impossible to tie in a new section of foundations to the original part securely. The dimensions of the building could be changed but this might affect other aspects of the design. Again, it would be expensive to go back to the designer to change this.

Not surprisingly, sometimes the walls are simply not built on the centre line of the foundations. Within reasonable tolerances this may have no effect.

If, however, the walls are built too far off the centre line, it can create a tendency for the foundations to tip over (Figure 3.5.1). This would rarely affect the whole building, as in most cases the bricklayer would reasonably try and achieve the best fit possible. If the line of the foundation was not straight, most people would take the best line they could, perhaps veering towards the internal face in one place and veering to the external face elsewhere. It is not difficult to imagine a situation where the foundations had been laid too short. In order to get the correct elevation length, the bricklayer would have to build up to the edge of the foundations at both ends.

Getting the foundations straight or to the correct length is not that difficult. By far the harder exercise, particularly on a sloping site, is getting the foundation plan square with 90° corners. To achieve a square building on top of out of square foundations, just one corner or just part of an elevation might be off centre.

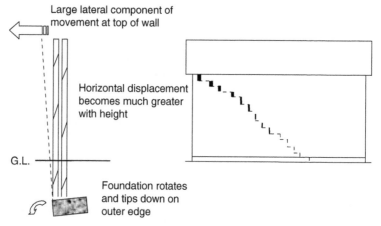

Figure 3.5.1   Rotational movement caused by eccentric loading.

The overall crack pattern looks very much the same as any other form of subsidence or settlement. The horizontal component of the movement is likely to be greater than would normally be seen with subsidence because the movement is rotational. The rotational movement is also likely to make the angle of the crack lower than the 45 degree angle, normally associated with subsidence.

In some cases, it may just be a contributing factor to another form of movement. For example, the foundations may have been built off centre, but under normal conditions it may not have any effect on the building. Imagine then, some decades later, when a nearby tree causes a little clay shrinkage. The clay shrinkage might not be sufficient on its own to cause subsidence. Just a little shrinkage however, on the outer edge of the house, could be exploited by the eccentric load of the footings. Together the two factors could cause movement.

This is probably more likely with trench-fill foundations. The ground shrinks from the surface down and, as it does so, a gap will appear between the face of concrete footing and the side of the trench. This will create a tapered void, wider at the top than the bottom. An eccentric force might then tip the footings into the tapered void.

The remedial action will be to underpin the wall with new foundations in the affected areas. This is a difficult and expensive operation. If the horizontal displacement is severe enough, then local rebuilding may be required.

3.6

Uneven Loading

Buildings are unevenly loaded, particularly where they change height, for example, from single-storey to two-storeys. Normal seasonal factors, such as moisture content changes in the subsoil, cause differential movement between the two parts. The cracks are cosmetic in nature. They are non-progressive but if re-pointed they will recur seasonally.

Key features

1. Small cracks, probably hairline to 1 mm width.
2. Generally even width although some minor rotation may occur.
3. Crack continues below damp-proof course but may become invisible.
4. Crack in line with change in load.

When the load distribution is uneven, for example when a building changes from a single storey to a two-storey height, a crack can occur on the intersection (Figure 3.6.1). On sand or gravel soils this can occur soon after construction. On clay soil it can develop gradually over years.

Seasonal changes in soil volume cause the building to lift and fall. It is much easier for expansion to lift the single storey part of the building than the two-storey part. A crack occurs roughly in line with the change in load point.

Practical Guide to Diagnosing Structural Movement in Buildings, Second Edition.
Malcolm Holland.
© 2023 John Wiley & Sons Ltd. Published 2023 by John Wiley & Sons Ltd.

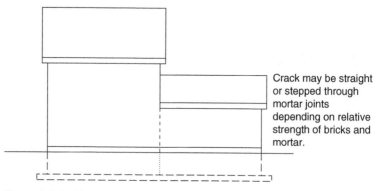

Crack may be straight or stepped through mortar joints depending on relative strength of bricks and mortar.

Figure 3.6.1   Uneven loading on foundations.

The crack may be straight through the bricks or blocks, or it may follow the mortar joints. This depends on the relative strength of the bricks and mortar.

The cracks are generally minor and uniform in width, although a little rotation may occur causing the crack to taper, being widest at the top and narrower at the bottom. As the crack is only likely to be hairline to 1 mm this will be minor. The crack will not be significant in structural terms and no remedial action is usually required. On exposed elevations however, a small crack might allow moisture penetration. In such circumstances the crack may need to be repaired or sealed for this reason.

Once a crack has occurred it creates a weak point. If it is re-pointed is likely to recur seasonally on the same line.

Load Concentrations on Foundations

Large openings in buildings create load concentrations at support points. These concentrated loads may not be able to spread out through the bond of the brickwork before reaching foundation depth. Localized overloading of foundations may occur.

Severity of the defect will depend on the facts in each case, but it is usually relatively minor.

Key features

1. Large openings.
2. Typically, hairline to 2–3 mm.
3. Rarely serious or progressive.
4. Recurs seasonally.

Where there are large openings in walls, the weight is concentrated down narrow sections of brickwork. This is most common where the opening is at the corner of a building, and the load is concentrated down a narrow pier. With a normal depth foundation, the load is often unable to spread out through the bond of the brickwork before it reaches the foundation. The result is an uneven load distribution. This will lead to differential

Practical Guide to Diagnosing Structural Movement in Buildings, Second Edition.
Malcolm Holland.
© 2023 John Wiley & Sons Ltd. Published 2023 by John Wiley & Sons Ltd.

compression of the subsoil and the foundation will settle more where the load is greatest (Figure 3.7.1).

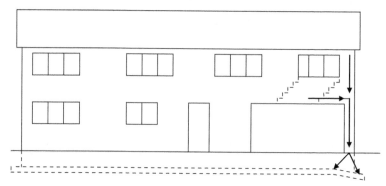

Load concentrated on pier is unable to
spread out before reaching foundation

Figure 3.7.1 Concentrated load on foundations.

This is rarely significant in structural terms. Differential movement is likely to create a crack from hairline to a couple of millimetres. It may; however, affect the function of secondary elements, for example if door openings move out of square, the doors may not fit and may not be able to be secured.

In severe cases the foundation may crack where it is bent. If the foundation cracks, then movement could be greater, but once the soil has been compressed enough to resist the load, the movement will cease. If cracks are re-pointed however, they will recur seasonally because of the uneven loading.

In severe cases the foundation could break, and this might allow the section of wall in question to become unstable. This would be relatively rare under normal conditions.

3.8

Differential Foundation Movement

Where foundations are set at different depths, they will be subject to different degrees of seasonal movement. This causes differential movement at the junction between the different parts. This commonly occurs when extensions are built onto old buildings. The old parts are likely to have relatively shallow foundations. The extension foundations have to be built to the depth required by modern regulations. The cracks are cosmetic in nature. They are non progressive but if re-pointed they will recur seasonally.

Key features

1. Different foundation depths.
2. Different foundation types.
3. Hairline to 2–3 mm usually.
4. Larger cracks if there are local influencing factors such as tree.
5. May be wider at top due to rotational component.
6. Recurs seasonally.

Practical Guide to Diagnosing Structural Movement in Buildings, Second Edition.
Malcolm Holland.
© 2023 John Wiley & Sons Ltd. Published 2023 by John Wiley & Sons Ltd.

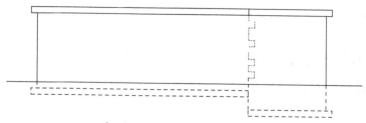

Crack may follow bond through the mortar joints or may be straight through the bricks and mortar depending on relative strength of bricks and mortar.

Figure 3.8.1   Differential foundation movement.

The seasonal variations of moisture content in the soil will affect the foundations. Such changes become less variable with depth. If the depths of the foundations vary under two parts of a building, the amount of seasonal movement will vary, and the two parts will move differentially to each other. Modern foundations usually have to be set at a greater depth than foundations in older buildings. Where a modern extension is added to an older building, the new part will most likely have foundations deeper than the original. All other things being equal, differential movement due to foundation depth is likely to occur.

Cracking is likely to be minor, from hairline to perhaps a couple of millimetres. Movement will not be progressive once equilibrium has been reached after construction. If the crack is re-pointed it will recur seasonally. The cracks may follow the bond of the brickwork through the mortar joints or run straight through the bricks and the mortar (Figure 3.8.1). This will depend on the relative strength of the mortar when compared to the bricks. Cracks may be larger if there are local factors to increase the potential for movement. For example, trees or shrubs nearby may not be big enough to cause serious failure of the foundations, but they will extract some moisture. Seasonal movement, exacerbated by nearby trees, will have the greatest influence where the foundations are shallow.

Cracks are likely to be reasonably uniform in width but can sometimes be wider at the top because of some rotational component of the movement. This occurs due to the shelter provided to the soil under the building. The soil around the building is exposed to the sun and rain. It will be subject to greater seasonal changes than the soil under it. The subsoil surrounding the building goes up and down to a greater degree than the soil under it. This causes the building to rotate or bend.

Figure 3.8.2 Porch on raft foundations. House on strip foundations.

In most cases no remedial action will be required unless the movement cracking is leading to other problems such as damp penetration.

Commonly very small extensions, such as porches for example, are built on shallow raft foundations which are not as deep as the original footings. In addition, raft foundations spread the load much more evenly than strip footings. This difference in loading, caused by the different foundation types, often leads to differential foundation movement. Figures 3.8.2 and 3.8.3 show a porch and bin store on a raft foundation connected to a house on strip foundations. The magnitude of movement in this example has been exacerbated by the proximity of shrubs and bushes in conjunction with a shrinkable clay soil.

Similarly, bay windows in older houses, also quite commonly have foundations which are not as deep as the main part of the house.

Such movement will also be quite commonly seen in factory buildings where the shed part of the factory has pad foundations to the columns, but the attached office part of the building has strip foundations to support the walls.

Figure 3.8.3   Close up of cracking between house and porch.
The magnitude has been exacerbated by clay shrinkage.

Figure 3.8.4   Differential movement where an extension joins to original.

3.9

Initial Settlement after Construction or Alterations

When a building is built it applies a load to the soil. This compresses the soil. The compression makes the soil stronger. When the soil has been compressed sufficiently, it becomes strong enough to support the building. During this process the building settles down until it reaches equilibrium. Some cracks will occur during this process. When an extension is added to an existing building it also undergoes an initial settling process. This can cause movement to occur between the original part and the new extension. After initial settlement the movement should stop. Cracking is usually minor. It is non progressive but if repointed it will recur seasonally.

Key features

1. Different loading.
2. Different foundations.
3. Different ages.
4. Variable movement hairline to 5 mm.
5. One large movement to new equilibrium.

Practical Guide to Diagnosing Structural Movement in Buildings, Second Edition.
Malcolm Holland.
© 2023 John Wiley & Sons Ltd. Published 2023 by John Wiley & Sons Ltd.

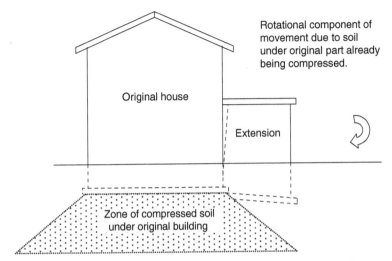

Figure 3.9.1 Initial settlement after alterations and zone of compression.

When a load is applied to a soil, it compresses it. As the soil is compressed the building will sink. As the soil is compressed, it gains strength. When the soil has been compressed sufficiently, to be strong enough to resist the weight of the building, the movement will stop. The building and the soil have then reached a state of equilibrium. Provided that the initial settlement is reasonably uniform, it will not result in any structurally significant movement. When an extension is added to a building, it also has to go through the initial settling down process. This causes it to crack or separate at the junction with the original. In a sandy or gravel soil, the movement takes place relatively quickly, within the first few months or year (Figure 3.9.1). The degree of movement can vary from hairline up to 3–5 mm typically, depending on subsoil variations.

On clay soil the movement is likely to occur over a longer time scale. Movement may occur over several years, although the main part of the movement will be within the first year.

Some rotational movement is not uncommon. This is because the soil under the original part of the building has already been compressed (See Zone of Compression in Figure 3.9.1). The soil under the building, which has already been compressed, will provide greater resistance to the new load than that further away, which has not been previously compressed. Hence the movement is greater in relation to the distance away from the original house, and rotation occurs as a result.

This type of movement can also happen to new buildings if the loads are uneven (see Chapter 3.6 Uneven Loading). For example, a building that is part three storey and part one storey, will apply an uneven load. The three-storey part will settle more than the one-storey part. If the subsoil is such, that a considerable amount of compression results, dissimilar movement will occur between the two different parts. Once the initial settlement has taken place a new equilibrium is reached. After that, no further movement would be expected, other than some normal seasonal movement.

Movement of this type is usually insignificant, being hairline to a couple of millimetres. Where the soil is perhaps a little loose, for example in gravels or sands, the degree of initial settlement may be greater and sometimes up to 3–5 mm or so.

Provided that equilibrium is achieved, no action is usually required, unless the movement is leading to other problems such as damp penetration.

3.10

Differential Foundation Settlement Cracking between Chimneys and Party Walls, in Alleys Running through Terraces

This defect is mostly found in 18th century, 19th century and early 20th-century terraced housing, mostly in urban areas. Alleys run through the terrace. Chimney stacks are more heavily loaded than the alley walls. The substantial weight causes differential settlement. Vertical shear cracks hairline to 1mm wide. Typically, movement takes place over the first few decades of the building's life. Movement is not progressive. No remedial action is usually needed.

Key Features

1. Terrace housing with access alleys through the terrace.
2. Chimneys built into the alley walls.
3. Vertical shear cracks typically, hairline to 1mm wide.
4. Vertical displacement from weight, differential loading.

In the UK it is common to find houses built as long terraces; particularly in towns, where large numbers of houses were needed to accommodate workers for industry and commerce. This is particularly common in the

Practical Guide to Diagnosing Structural Movement in Buildings, Second Edition.
Malcolm Holland.
© 2023 John Wiley & Sons Ltd. Published 2023 by John Wiley & Sons Ltd.

later part of the 18th century, throughout the 19th century and early part of the 20th century, prior to World War One (1914). During this period, houses were heated by open solid fuel fires (coal in urban areas).

When long terraces of housing are built, one of the technical problems to overcome is how to gain access to the rear, without going through the house. Sometimes, access was made by rear service roads or rear alleys. An alternative was to run alleys through the terraces at ground floor level, beneath the first-floor bedrooms.

When an alley is run through a terrace, the designer has several options about how to divide the properties at the party wall position. It is common to find an alley running through the terrace with the fireplaces and chimneys built into the party wall forming the side of the alley (See Figures 3.10.1 and 3.10.2).

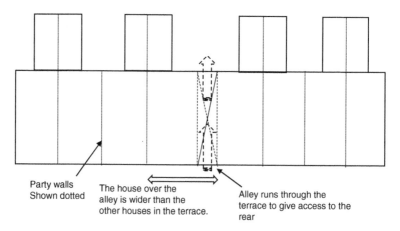

Party walls
Shown dotted

The house over the
alley is wider than the
other houses in the terrace.

Alley runs through the
terrace to give access to the
rear

Figure 3.10.1   Plan view of terrace of houses with alley running through the terrace.

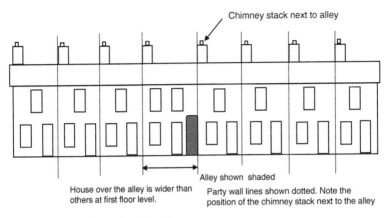

Figure 3.10.2 Front elevation of terrace.

With this design it is common to find vertical cracks on the line of the chimney, within the alley (See Figures 3.10.3 and 3.10.4).

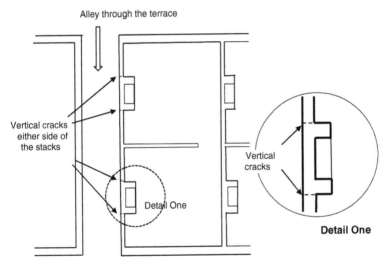

Figure 3.10.3 Plan position of cracks either side of the chimney breasts.

The cracks are caused by the difference in weight between the chimney breasts and the walls. The alley walls will be 225mm thick, at most (the thickness of one brick length). It is, however, common to find that the alley walls are only half a brick thick (110mm), either side of the chimney breast. Thin walls are particularly found in low-quality artisan housing of

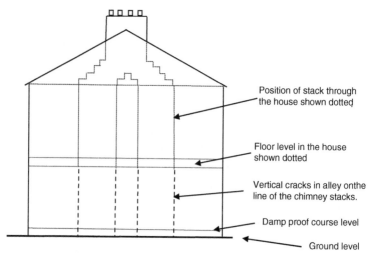

Figure 3.10.4   Elevation position of cracks either side of the chimney breasts.

the period. Cracking is more likely if the walls are only 110mm thick, because of the difference in weight between the thin walls and the chimney stacks.

The chimney breast will be 400–500mm thick. This mass of brickwork is very heavy. As the walls settle down naturally after construction, the weight of the chimney breast section causes it to settle into the ground slightly more than the thinner sections of the walls. A vertical shear crack occurs.

The cracks will be vertical from ground level, all the way to the visible top of the wall in the alley. The cracks will be uniform in width. The width of the cracks will be small, hairline to 1mm wide. The cracks will go through the damp-proof course and to ground level. The cracks will continue below ground level to the foundations.

A close inspection of the horizontal mortar joints may show a slight step up/down either side of the junction of the crack. This vertical shear displacement is, however, often so small that it is difficult to determine. This is particularly the case if the brickwork is a little irregular.

The wall would have cracked internally as well but the internal crack would have been filled and decorated numerous times over the lifetime of the building and there is unlikely to be any evidence of it.

This form of cracking will be minor and historic in nature. No remedial action is required.

If the crack is re-pointed on the alley side, it is likely to occur again as a hairline crack. That is simply due to differential movement, daily or seasonally.

3.11

Leaking Drains and Water Discharge near to Buildings

> *Rainwater discharged adjacent to a building can cause localized subsidence. Broken drains underground can cause localized subsidence. Rainwater should be discharged away from the walls of a building. Any under ground drainage leaks need to be repaired. The severity of the defect is dependent on the facts and degree of movement in each case.*

If a significant volume of water is discharged near to a building, either above or below ground, it can cause foundation movement (Figures 3.11.1 and 3.11.2). Cohesive soil (clay) will become softer and more plastic if the moisture content is increased. Non cohesive soil (sands and gravel) may lose the fine particles to water running through it, causing it to subside locally.

Key features

1. Localized defect adjacent to location of defective drain (above or below ground).
2. 45° angle of crack.
3. Tapered crack getting wider with height.
4. Slow and progressive.
5. Movement may increase seasonally with wet weather.

Practical Guide to Diagnosing Structural Movement in Buildings, Second Edition. Malcolm Holland.
© 2023 John Wiley & Sons Ltd. Published 2023 by John Wiley & Sons Ltd.

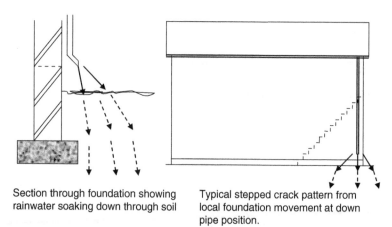

Section through foundation showing rainwater soaking down through soil

Typical stepped crack pattern from local foundation movement at down pipe position.

Figure 3.11.1 Downpipe discharging by foundations and associated crack pattern.

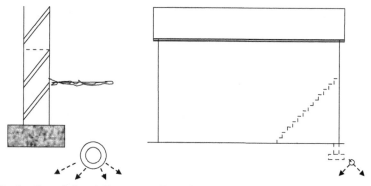

Section through foundation showing leaking underground drain

Typical stepped crack pattern from local foundation movement at corner by leaking drain

Figure 3.11.2 Leaking underground drain by foundation and associated crack pattern.

There are a number of reasons why movement of this type could occur. If there is no provision for underground drainage, and rainwater is discharged adjacent to the wall of a building, it can cause damage. The risk of this occurring is greater if the foundations are shallow. If an underground drain breaks, water can be released below ground where it may go un-noticed for a long time.

The crack pattern associated with this type of local subsidence is the typical roughly 45° crack, through the bond of the brickwork. Typical of a subsidence crack, it will be tapered, becoming wider with height, due to the effect of rotation. Movement is likely to be slow and progressive. Movement is likely to be seasonal, associated with the wetter parts of the year.

Movement can be stopped by improving the drainage. This could simply be piping the water a few metres away from the wall. Alternatively, it might involve the construction of an underground drainage system, discharging to a drain, watercourse or soakaway. Individual circumstances might dictate the possible solution. In a rural area it might be feasible to discharge the water onto the ground away from the building. The same solution might not be possible in a more urban area.

If the problem is a leak in an existing underground drain, this would have to be traced and repaired. The drains can be inspected underground using a closed circuit television camera, pushed down the drain from the nearest inspection chamber access point.

After the cause of the movement has been identified and repaired, the ground should be allowed to dry out. This could take several months or a year.

The amount of repair work required to the wall will depend on how much movement has taken place. If the movement is minor, then re-pointing may be sufficient. If the movement has broken the bricks in the wall, new brickwork can be stitched in. If the movement has been allowed to progress to a severe stage, localized re-building of the wall may be required, possibly on new foundations.

3.12

Drains and Drain Trenches

Modern regulations require drain trenches to be back filled with concrete, when they are laid within the influence of the foundations of a building. Historically this was not the case. Old drain trenches would just be filled with selected fill or gravel. Foundation movement can occur, as pressure from the footings presses into the loose backfill. This may just result in some initial movement, which then ceases. If the initial movement breaks the drain, progressive long-term movement can occur. Defective drainage needs to be repaired or re-laid. The severity of the defect is dependent on the facts and degree of movement in each case.

Key features

1. Common in old houses.
2. Drains deeper than foundations.
3. Initial large settlement.
4. Possible drain damage and progressive movement.
5. Wall may lean.

Practical Guide to Diagnosing Structural Movement in Buildings, Second Edition.
Malcolm Holland.
© 2023 John Wiley & Sons Ltd. Published 2023 by John Wiley & Sons Ltd.

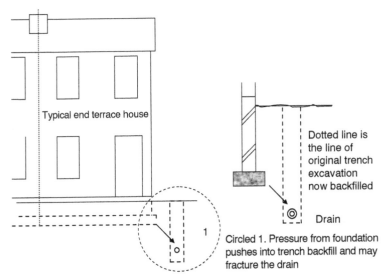

Typical end terrace house

Dotted line is the line of original trench excavation now backfilled

Drain

Circled 1. Pressure from foundation pushes into trench backfill and may fracture the drain

Figure 3.12.1 Drain trench within zone of foundation influence.

In modern construction, any drain laid at a depth below adjacent foundations and within the 45° angle of influence, must be bedded in concrete. The trench in which it is laid must also be filled with concrete to a level above the angle of influence. This measure is intended to stop pressure from the foundation, pushing into the side of the backfilled trench. If the trench was simply backfilled with shingle or selected fill, pressure from the foundation would compress the fill from the side. Historically this was not a requirement.

It is not uncommon to find buildings, where the drain runs are below the level of the foundations. This is particularly common in pre–World War One buildings, but also in interwar buildings. It is very common in terraced houses, where there are long drain runs along the length of the terrace at the rear. Almost inevitably, the slope of the pipe over such a long run, results in the depth being below foundation level. Foundations in houses of this period are likely to be shallow. The pipe will then usually turn through 90° and run down the side of the end terrace property, to connect into the sewer in the road. The drainpipes in properties of this age would usually be clay with rigid joints. The back fill would be, at best, selected fill with the larger of any stones or old bricks removed. The pressure from the foundations can push into the relatively loose fill, allowing both settlement and outward slip (Figure 3.12.1).

The flank wall of an old terrace house is quite typically poorly restrained. Most commonly floor joists run front to rear in terrace housing and chimney breasts usually form part of the party wall (see Chapter 2.7 Lack of Lateral Stability). As a result, it is quite common for the wall to move out of vertical, rotating outwards at the top.

The initial movement can be quite pronounced. If the drain is not fractured by this movement, then a new equilibrium may be reached, and the movement may cease. Evidence of this historic initial movement can often be seen in Victorian Houses.

If on the other hand the drain is fractured, the initial movement can be followed by long-term progressive movement, as the leaking drain weakens the soil.

With a clay subsoil, the increased moisture content increases the plasticity of the soil. This causes it to soften, leading to compression. In sandy soils, leaking drains may wash away fine particles. This leaves voids. The larger particles slip and compact into the voids, leading to settlement in a building. Any leaking drain within close proximity can have this effect on a building (see Chapter 3.11 Leaking Drains and Water Discharge Near to Buildings).

In such circumstances, where drain problems are suspected, the drains should be tested and repaired or replaced as necessary, before carrying out other remedial work. If caught early enough, remedial work to the walls may merely be cosmetic. In more severe cases, lateral restraint may need to be improved by tie bars or straps. Underpinning of the foundations may also be required. In the most severe cases, where an adjacent wall has moved out of vertical to a large degree, re-building may be the only option. In such cases it is a matter of degree based on the circumstances and severity in each case.

Cracking Associated with Raft Foundations

Raft foundations are sometimes used instead of traditional strip foundations. This is usually because of poor ground condition. Also used with lightly loaded buildings particularly garages where the raft can also be the floor. Wall loads are concentrated on the perimeter of the slab. There is differential expansion movement between the slab and the brickwork. Long-term creep can cause the raft to bow. Movement can be expansion cracking, bowing or both combined. Cracks can be confused with subsidence. It is therefore essential to be able to identify the construction and characteristics of the cracking. Cracking is generally small and not progressive.

Key Notes

1. Reasons for Raft foundations.
2. Weak soil – poor bearing capacity.
3. Shrinkable soil.
4. Made up ground.
5. Typically, bungalows and garages.
6. Expansion cracking is vertical and even.
7. Expansion and bowing combined is wider at the top.
8. Bowing and weak routes through the brickwork can cause cracks to stagger.
9. Identify the construction by visible concrete base at perimeters.

Practical Guide to Diagnosing Structural Movement in Buildings, Second Edition. Malcolm Holland.
© 2023 John Wiley & Sons Ltd. Published 2023 by John Wiley & Sons Ltd.

Most buildings, which are built with load bearing walls, are supported on strip foundations, running under-ground the full length of the wall. Larger framed buildings, with columns and beam frames, have pad foundations under each column. This of course assumes reasonably good or at least average ground conditions. Sometimes specialist foundations such as piles have to be used, for example, either to make the foundations deeper to avoid shrinkage of the soil or to achieve greater load-bearing capacity. Another type of foundation design that is sometimes used is a raft foundation. There are several reasons why a raft foundation might be chosen.

A raft foundation may be used on soils that swell or shrink. This might be a clay soil with trees nearby, where there could be seasonal movement. It could be where trees have been removed before construction, and some swelling up of the ground (heave) is anticipated.

Raft foundations, as the name suggests, are a raft of concrete that 'floats' on the ground, like a wooden raft might on the sea. In this context, the term 'floating' just means laying on and not connected. It does not mean hovering.

The raft may go up or down a little seasonally. The raft may tip a little. The point is, that the foundation moves as one unit, and the building moves with it. To be able to do this, the concrete raft has to be strong enough in tension to be able to move as one. Concrete on its own does not have the tensile strength to achieve this, and therefore concrete raft foundations are usually reinforced with steel mesh. In some cases, where repeated movement is not anticipated, and where the loads are very low, a simple unreinforced concrete slab might be acceptable. For example, precast concrete panel garages sometimes have an unreinforced slab which forms the garage floor and a solid base for the wall panels. These buildings are low value and will also allow a little movement at the panel joints. Timber frame huts would also be another instance where an unreinforced slab might work adequately.

Raft foundations are often suited to low-weight buildings, such as garages and bungalows. If the loads are low, the foundations would not have to go so deep to find a good load-bearing stratum. Strip foundations, however, must be a metre deep on clay soils to prevent seasonal movement. If the foundations are one metre deep to avoid seasonal movement, they might be on a soil that is many times stronger than it needs to be in relation to the loads supported on it. An alternative is to build a shallow raft that also forms part of the floor structure.

Raft foundations are often used when the sub soil has poor bearing capacity. They are also used on ground that might have uneven bearing capacity, such as made-up ground. The large area of a raft can spread the load. It can bridge over weaker soils until they are all compressed together under the load.

Raft foundations spread the load of the building over a wide area. Inevitably, however, buildings have walls which concentrate the loads in certain areas and particularly around the perimeter. Raft foundations must be designed to support the concentrated loads and be strong enough in these areas to be able to spread the loads out.

Consequently, raft foundations are usually thickened with a 'down stand' around the perimeters to support the loads (see Figure 3.13.1).

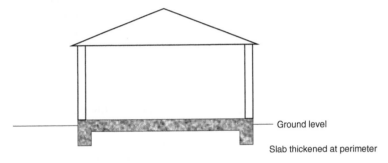

Ground level

Slab thickened at perimeter

Figure 3.13.1 Concrete raft with perimeter down stand.

An alternative is to extend the rafter beyond the walls (see Figure 3.13.2). This design solution is less common.

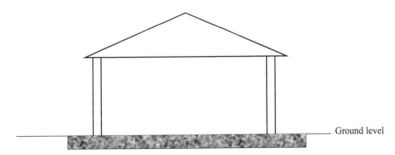

Ground level

Figure 3.13.2 Concrete raft extending beyond walls.

In both forms of construction, the raft is visible around the perimeter, at or above ground level and beneath the brickwork. This is how a raft foundation can be identified. If the brickwork extends beneath the ground level, it is a strip foundation. If the concrete is visible beneath the brickwork, it is a raft.

One other limiting factor on the use of rafts, is size. The slab has to be cast within form work and tamped down level. There is a practical limit to how long a tamping plank can be, and still be handled by the workforce.

Another limiting factor is the nature of the chemical reaction that sets the concrete. When concrete sets, there is a chemical reaction in the setting process that gives off heat. Concrete expands during this reaction and then contracts as the reaction completes. This expansion and contraction can cause concrete to crack if it is cast in large areas. Cast concrete bays are therefore usually about 6 metres by 4.5 metres max. Each bay must be connected to the next bay by an expansion/contraction joint. Dowels must be built in between the bays to give continuity between bays.

Expansion joints in the floors would not be acceptable in most buildings, such as homes or offices. Some form of 'floating' floor must be built over the raft to cover this. In this context the term 'floating' means separated and resting on by gravity, not hovering in air.

One of the most common situations where raft foundations are used, is the construction of garages. In garage construction, the raft can form the foundation and the slab. Garages for cars do not have to be insulated and therefore the walls are usually thin and relatively light weight. A single garage is usually about, or a little less than, 6 metres by 3 metres. A double garage can be built with two connected bays with an expansion joint down the middle. A terrace of garages can be cast in connected bays at every garage party wall.

Sometimes in garage terraces, the raft is formed with no expansion/contraction joints. In effect the raft is left to move and form its own expansion joints by cracking. If the steel mesh is continuous, the cracking tends to form uniformly, and reasonably parallel to, the party walls, which act like restraints. The effect is similar to tearing paper on a fold line. The cracks are usually just a millimetre or two, and in the context of a garage, have no real effect on the performance of the building.

Concrete has a higher coefficient of expansion than brickwork. Clay brickwork needs an expansion joint around every ten metres. Concrete needs an expansion joint about every six metres. Consequently, it is common to find expansion cracking in buildings built on raft foundations. Expansion cracks tend to be vertical and even in width, please refer to Chapter 2.1 'Expansion Cracking'.

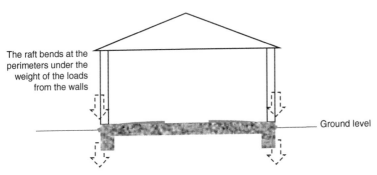

The raft bends at the perimeters under the weight of the loads from the walls

Ground level

Figure 3.13.3   Raft bending under load.

Raft foundations, however, often also suffer from bending under load with time. See Figure 3.13.3.

Reinforced concrete is a fairly rigid and permanent material. Although it is fairly rigid, it is elastic to some degree and will 'creep' under load and over time. The weight of a building is applied to the raft under the perimeter load bearing walls and also onto any internal load bearing walls. The perimeter walls are usually heavier because they are thicker, and often because they carry more of the roof loads and any floor loads. Over decades, the raft tends to bend around the perimeters and becomes 'domed'. This would not be visible to the naked eye. It would be so small that it might even be difficult to measure with a simple bricklayer's spirit level. Quite accurate levelling equipment, such as an optical level or laser level, might be needed.

The bending of the raft causes the wall to open out as well (see Figure 3.13.4). This can be sufficient to cause cracks in the brickwork. As the angle of displacement causes more movement at the top than the bottom, the cracks will be wider at the top. Tapered cracks, which are wider at the top than the bottom, are a characteristic commonly associated with subsidence. It is easy to jump to the wrong conclusion if the construction is not correctly identified.

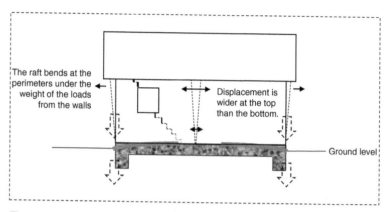

Figure 3.13.4 Bending of the slab makes cracking wider at the top. Cracks may also be staggered, following weak routes, for example through window openings.

The cracks will often exploit weak routes through window openings. This can also make the crack run diagonally, which is again a feature commonly associated with subsidence. It is essential to identify any distortion of the crack caused by the positioning of openings. Please refer to Chapter 1.5 Weak Routes.

The cracks are unlikely to extend through the concrete slab, as this will have bent by long-term creep, and being at the base of the wall, the opening effect will be very small. There is, however, the possibility that the cracking will coincide with an expansion/contraction crack in the slab.

In most cases, cracking caused by this mechanism is likely to be relatively small. It is likely to start as hairline at the bottom and widen progressively to probably 1–2mm or 2–3mm at the top. If the wall is a cavity wall, the cracking will be in both leaves of the wall.

Internally however, any evidence of cracking may have been covered during previous redecoration cycles. This is likely to be the case for habitable buildings such as bungalows, but not in garages.

The appearance of the cracks can be more pronounced if both expansion and bending combine. This is not uncommon in a long terrace of garages or even with semi-detached double garages. Historically, garages were often built without any expansion joints.

The typical vertical pattern of expansion cracking can be distorted out of vertical by the bending of the slab. Inevitably, not all mortar joints have the same strength. A combination of expansion and bending may take the crack off at a slight angle. This is again a feature that can be mistaken for subsidence.

The difference, however, is that only the horizontal displacement gets greater with height. The horizontal cracking on the bed joint will remain hairline. Whereas with subsidence, the crack would display both vertical and horizontal displacement. By understanding the construction and mechanism, as described above, this confusion can be prevented.

In most cases the cracks will be so small that no remedial action will be required, other than for cosmetic reasons.

If expansion cracks are simply re-pointed, they will crack again during the next heating cycle. Expansion joints could be cut and formed in the brickwork. Such work should not be undertaken without proper professional advice to ensure lateral stability is not affected.

There may be a few rare cases where the raft is completely overloaded, to the point where it has fractured and become detached or hanging on the reinforcing. In such a case, the movement would be more pronounced, at possibly around 5–10mm. This would be a severe failure and expensive underpinning repairs may be required. See Chapter 4.10 Underpinning p. 259.

The crack is just to the right of the fan trellis, approximately central on the elevation. Highlighted on the photograph. Note that this is not the actual crack. It has been drawn on to show its position. See Figure 3.13.6 for close up of the crack.

Figure 3.13.5   Double back-to-back garage block built off a raft foundation.

Figure 3.13.6   Expansion cracks in the double garage, with all displacement horizontal, side to side. The vertical component of the cracking is wider at the top because the raft has bent. The cracking on the horizontal bed joint remains hairline. The crack is wide enough to be visible at upper level but then tapers away to hairline and invisible on the photograph at lower level.

Figure 3.13.7   A concrete raft to a terrace of three single garages, identified by the concrete raft visible below the brickwork at ground level.

Figure 3.13.8 Cracking centrally on the terrace of three garages. All the displacement is horizontal, side to side. The vertical component of the cracking is wider at the top because the raft has bent. The cracking on the horizontal bed joint remains hairline. The crack is 2mm wide at the top but then tapers away to hairline and invisible on the photograph at lower level.

Part 4

Repair Methods

4.1

Introduction

> *The severity of the defect will determine the type of repair required. Many defects will become progressively worse. Early diagnosis may prevent damage from developing to a serious level. In each case it is a matter of degree and judgement. This is an overview of the techniques commonly used to repair buildings that have suffered from movement and cracking. These range from simple cosmetic re-pointing to underpinning foundations.*

The severity and nature of the defect will determine what type of repair is required. Many defects have the potential to become progressively worse with time. If they can be identified quickly, then the cause of the defect can sometimes be removed before serious damage occurs. Subsidence related to clay shrinkage and trees is perhaps the most obvious example of this.

When clay shrinkage first affects a building, the cracks start from nothing. If they are identified when they are hairline or just 1–2 mm, it might be possible to remove the tree or trees that are causing the problem, before it develops further. Once the trees have been removed, the soil regains its moisture content and reaches a new equilibrium. During this time some recovery is likely as the ground swells up again. In a case like this, perhaps no repair, or just minor re-pointing, would be required.

Practical Guide to Diagnosing Structural Movement in Buildings, Second Edition.
Malcolm Holland.
© 2023 John Wiley & Sons Ltd. Published 2023 by John Wiley & Sons Ltd.

If the same defect was not diagnosed, and was allowed to progress, the worst-case scenario would be very expensive underpinning repairs, or possibly even re-building.

With other defects, such as roof spread or lateral instability, the walls will move progressively out of vertical. If the defect is identified before the walls are significantly out of vertical, quite simple repairs can sometimes be carried out. It might be possible to simply improve the roof frame or improve restraint to the walls. If the walls are allowed to move too far out of vertical, complete demolition and rebuilding may be required.

In each case, it is a matter of degree and judgement. The specification of repair work, other than simple re-pointing, is perhaps best left to professionals such as Chartered Building Surveyors, Chartered Engineers or specialist firms. There are a number of specialist techniques designed specifically for the repair of building defects, such as the ones described in this book.

This section is intended as an overview of the common types of repairs that can be carried out, not a detailed specification.

Re-pointing

> *Re-pointing a crack is the most basic level of repair that can be car-*
> *ried out after the cause of the defect has been resolved. Re-pointing is*
> *often carried out to a poor standard with an inappropriate specifica-*
> *tion. There will always be difficulty matching re-pointing to the original.*
> *Where cracks are small it can be better to just leave them to avoid obvi-*
> *ous re-pointing.*

Re-pointing brickwork externally and filling cracks internally, is the basic level of repair that can be carried out when the damage is minor and the cause of the defect has been resolved.

When brickwork is re-pointed, the joint should be raked out to a depth of around 25 mm. In practice, unless the work is specified and supervised by a professional, most re-pointing will not have been carried out in this manner. In most cases re-pointing is carried out by just scraping away the surface of the mortar joint and applying a smear of mortar, just a couple, or a few millimetres thick.

It is also very common to find that a very hard cement-rich mortar has been used, in the mistaken belief that the resulting hard mortar will be more resilient against future movement. It is not surprising then that many previously re-pointed repairs will quickly show signs of cracking

Practical Guide to Diagnosing Structural Movement in Buildings, Second Edition.
Malcolm Holland.
© 2023 John Wiley & Sons Ltd. Published 2023 by John Wiley & Sons Ltd.

along the repair. This is commonly caused by shrinkage of the hard cement-rich mortar and normal seasonal flexing along the joint.

Mortar used in re-pointing should ideally match the appearance and strength of the original. Matching the colour will always be a problem. As near as possible is all that can be hoped. Only time and weathering will serve to blend the colour in. The pointing should be finished in the same way as the original. For example, if the original joints are finished with bucket handle pointing, the re-pointing should also be finished in this manner.

If the bricks are worn at the edges however, it is sometimes best not to bring the mortar right out to the face. Otherwise, the width of the bed joint will taper, in and out, depending on how each brick is worn.

Re-pointing cracks often results in a less pleasing appearance than just leaving the cracks. If the crack is less than a millimetre or two wide, and is not allowing other defects, such as water penetration, if may be better to leave it alone.

4.3

Re-pointing with Epoxy Mortar

> *When a crack is re-pointed it will often crack again on the same line. This is because the re-pointed joint creates a weak point. Re-pointing with an epoxy mortar increases the strength of the joint making it stronger than the original. These are proprietary products often applied by contractors who specialise in brickwork repairs.*

When a crack has been re-pointed it will nearly always crack again along the same point. The re-pointed crack is a weak point through the wall and will be exploited by normal seasonal movement. The only way to prevent this is to make the repair stronger than the original construction. Mortar is a brittle material, weak in tension. One way to improve the tensile strength of the mortar repair is to use an epoxy mortar. These are proprietary products and are often applied by contractors who specialise in brickwork repairs. Re-pointing with an epoxy mortar is a similar process to normal re-pointing. The depth of re-pointing can be varied to the extent of completely raking out the old joint.

Practical Guide to Diagnosing Structural Movement in Buildings, Second Edition.
Malcolm Holland.
© 2023 John Wiley & Sons Ltd. Published 2023 by John Wiley & Sons Ltd.

4.4

Stitching in Brickwork

> *When cracking has damaged the bricks, or the degree of movement is severe, it is sometimes necessary to cut out bricks and stitch in new bricks. It is difficult or impossible to provide an exact match. Repairs may be less aesthetically pleasing than just accepting some cracked bricks.*

When cracking has damaged the brickwork, or the degree of movement is more severe, it is sometimes necessary to cut out brickwork across the position of the crack. Again, it will be impossible to fully match any replacement bricks, so unless it is really necessary, it might be better to just accept some damaged bricks. Once new bricks have been stitched back in using an epoxy mortar, this strip of new brickwork will be stronger that the original. It is therefore possible that normal seasonal movement will cause hairline cracks either side of the repair.

Practical Guide to Diagnosing Structural Movement in Buildings, Second Edition.
Malcolm Holland.
© 2023 John Wiley & Sons Ltd. Published 2023 by John Wiley & Sons Ltd.

4.5

Reinforcing Brick Mortar Joints

> *Mortar joints can be reinforced with metal mesh or bars, set in epoxy mortar. This imparts tensile strength to the brickwork. Reinforcing can be designed to act as a lintel over openings, such as windows.*

Brick mortar joints can be reinforced by, raking them out, inserting metal mesh, metal strips or helical twist metal bars into the joints, and then re-pointing. These are proprietary products and repairs are often carried out by companies that specialise in this type of work.

The metal reinforcing provides tensile strength across the repaired area. This makes it stronger than the original. There is the possibility that minor cracks may develop either side of the repaired areas, where the reinforced brickwork meets the original. Where the reinforcing is inserted into mortar joints over openings, it can serve to provide enough tensile strength to act as a lintel (see Figure 4.5.1). In Figure 4.5.2, the re-pointed crack is visible, as the epoxy mortar is not an exact match. The horizontal lines, where the helical twist bars have been inserted, are also visible. Although this recent mortar repair is clearly visible now, it will "weather in" with time and become less obvious.

Practical Guide to Diagnosing Structural Movement in Buildings, Second Edition.
Malcolm Holland.
© 2023 John Wiley & Sons Ltd. Published 2023 by John Wiley & Sons Ltd.

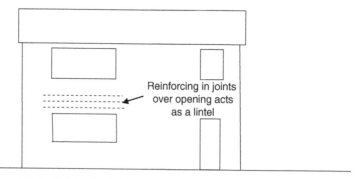

Reinforcing in joints
over opening acts
as a lintel

Figure 4.5.1   Reinforcing brick mortar joints.

Figure 4.5.2   Subsidence repairs to brickwork with helical bars and epoxy mortar. Horizontal re-pointed lines are visible every four courses where the bars have been inserted and the stepped line of the subsidence crack can be seen.

4.6

Tie Bars

Traditionally metal bars were inserted through buildings to improve lateral stability or to prevent roof spread. Where there are rooms within a roof it is often not possible to insert bars at their ideal structural position. The best position, from a structural point of view, could possibly interfere with the practical use of the room.

Traditionally, additional restraint was provided to buildings by using tie bars. This was often carried out to combat lateral instability or roof spread. The bars are commonly steel or iron, and about 25 mm in diameter. They are capped at either end with a metal plate in order to spread pressure on the walls. The plates are commonly round shaped, cross shaped, 'X' shaped or 'S' shaped (see Figure 4.6.2 and Figure 4.6.3). The bars are threaded through the building from one face to the other, usually just below the level of the ceiling or through the depth of the floor (Figure 4.6.1). The bars are usually finished with a screw thread at the ends. The end caps are tightened up on the screw threads. Alternatively, a screw coupling is provided in the middle for tightening. When a building is suffering from roof spread, the ideal position for the tie bar would be at the top of the wall. This would, however, be impractical in most cases, as it would be at waist or shoulder height within the first-floor room.

Practical Guide to Diagnosing Structural Movement in Buildings, Second Edition. Malcolm Holland.
© 2023 John Wiley & Sons Ltd. Published 2023 by John Wiley & Sons Ltd.

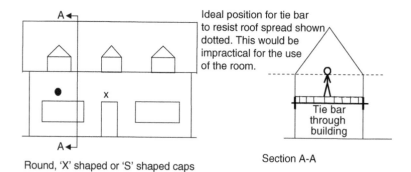

Round, 'X' shaped or 'S' shaped caps

Section A-A

Figure 4.6.1   Position of tie bars through a building.

Figure 4.6.2   Typical round capped tie bars at first floor and roof wall plate level.

Figure 4.6.3   Cross shaped tie bar cap.

Although in the examples above the tie bars are shown through the building from front wall to rear wall, they are also sometimes fixed to the floors or roof/ceiling joists, in order to provide restraint to flank walls in end terrace or semi-detached buildings.

4.7

Restraint Straps

> *Lateral restraint is provided in modern buildings by strapping floors and roofs to the walls, using light-weight steel straps. These can also be retrofitted into existing buildings, to improve lateral stability.*

Another method of providing lateral restraint is by using tie straps. Since the mid 1970s it has been a Building Regulations requirement to connect external walls to the floor, ceiling and roof structure. Where the joists/rafters run parallel to the walls this is achieved by building in restraint straps (Figure 4.7.1). They should be provided at a maximum of 2 metre centres. The straps are made from galvanized steel, 5 mm thick and 30 mm wide. The straps have to be built into the wall during construction and span across the adjacent three roof trusses or joists. They will normally be about 1200 mm–1800 mm long.

Buildings built before this time will not have this connection between walls and floors/roofs that span parallel to them.

Straps can be retrofitted between the walls and the floors/roof timbers. Bricks or blocks can be cut out of the walls. The straps can be cast into the walls with cast in situ concrete. Alternatively, they can be mechanically fixed, although this is unlikely to be as effective.

Practical Guide to Diagnosing Structural Movement in Buildings, Second Edition.
Malcolm Holland.
© 2023 John Wiley & Sons Ltd. Published 2023 by John Wiley & Sons Ltd.

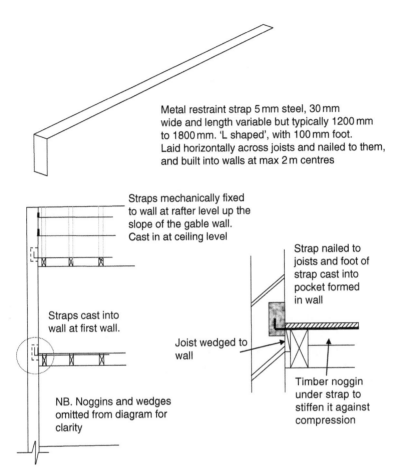

Metal restraint strap 5 mm steel, 30 mm wide and length variable but typically 1200 mm to 1800 mm. 'L shaped', with 100 mm foot. Laid horizontally across joists and nailed to them, and built into walls at max 2 m centres

Straps mechanically fixed to wall at rafter level up the slope of the gable wall. Cast in at ceiling level

Strap nailed to joists and foot of strap cast into pocket formed in wall

Straps cast into wall at first wall.

Joist wedged to wall

NB. Noggins and wedges omitted from diagram for clarity

Timber noggin under strap to stiffen it against compression

Figure 4.7.1 Restraint straps.

4.8

Buttresses/Piers

> *Buttresses or piers can be built into walls to provide lateral stability or resist roof spread. This was a common traditional solution.*

Another way to provide lateral restraint to walls is to build in buttresses or brick piers. This is not uncommon in older houses and is often used to combat roof spread. Sometimes the buttressing is disguised by forming it into a porch (Figure 4.8.1). Sometimes a porch provides a buttressing function by chance rather than by design.

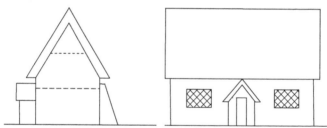

Typical example of a thatch roof cottage buttressed by raking brick buttress at the rear and by a porch built on the front. The dotted lines show the position of floor and room within the roof.

Figure 4.8.1 Buttressing to external walls.

Practical Guide to Diagnosing Structural Movement in Buildings, Second Edition.
Malcolm Holland.
© 2023 John Wiley & Sons Ltd. Published 2023 by John Wiley & Sons Ltd.

Preventing Roof Spread

Roof spread can be resisted by a number of methods. The design of the roof can be improved in some circumstances. Tie bars can be introduced. Walls can be buttressed to resist roof spread. Alternatively, a ridge beam can be used. In theory a ridge beam is an attractive alternative, as it reverses the forces involved. The roof will then pull in, rather than push out. In practice this can rarely be achieved. It is usually impractical to bring a large enough beam into an existing building.

Roof spread is a specific defect that can be prevented or mitigated, by a number of different methods. The outward push of the roof on the walls can be resisted by normal lateral restraint methods, such as tie bars or buttressing. The most common obstacle to this approach occurs when the space within the roof is needed to be used as a room. This often prevents the introduction of tie bars through the building for practical reasons. The room in the roof could not be used if a metal bar passed through it at shoulder height.

An alternative approach is to change the design of the roof or to strengthen the roof in order to prevent it from pushing the walls outwards.

The most elegant solution to roof spread would in theory be to insert a strong beam under the ridge of the roof. This holds the ridge up. By doing

Practical Guide to Diagnosing Structural Movement in Buildings, Second Edition.
Malcolm Holland.
© 2023 John Wiley & Sons Ltd. Published 2023 by John Wiley & Sons Ltd.

so, it not only prevents further spread, but it reverses the forces, and the roof pressures would then pull the walls in. This reverses the roof spread. Think of this like a book (Figure 4.9.1).

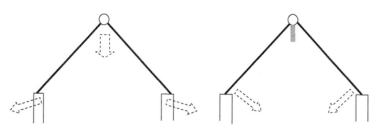

Imagine a ring bound book placed over two supports like a roof. The self weight of the book would cause it to push the supports out and the apex would drop.

Imagine now putting a ruler through the book, underneath the ridge; and holding the ruler up so the ridge could not drop. The book would then hinge at the apex and the weight of the leaves would try to drop in back together.

Figure 4.9.1   Explanation of roof ridge beam action.

In practice this can rarely be achieved. In most cases it would be too difficult to get such a long beam into an existing building. The beam would also have to be very strong or deep to resist deflection at mid span. It would also require very strong support points as it would have to support the major part of the roof load.

There are some other methods to strengthen and stiffen a roof in order to provide increased resistance against outward thrust. For example, strengthening the purlins or putting in new purlins to reduce deflection.

4.10

Underpinning

If the foundations of a building have suffered movement and the cause of the movement cannot be removed, the foundations may have to be improved. The foundations are underpinned to increase their depth. Traditional underpinning involves digging pits beneath the existing foundations and then creating a new foundation underneath.

A number of specialist mini piling techniques have been developed to underpin a building.

Underpinning is expensive and disruptive.

If movement in a building is foundation related, and the cause of the movement cannot be removed, underpinning to improve the foundations can be carried out. A common example of this is where a building has been affected by subsidence, caused by clay shrinkage and trees. There are several reasons why a tree might be more important than a building. Trees can take hundreds of years to reach their magnificence and are relatively unique. Their value to the history of a location and the environment can far outweigh the value of one specific building. Many trees are protected by preservation orders or conservation areas for this very reason.

If the tree cannot be removed, the risk of future movement will remain, unless the foundations are taken to a deeper depth. Increasing the depth

Practical Guide to Diagnosing Structural Movement in Buildings, Second Edition. Malcolm Holland.
© 2023 John Wiley & Sons Ltd. Published 2023 by John Wiley & Sons Ltd.

of foundations, or building new foundations under a building, is known as underpinning.

Underpinning is however, a very disruptive and expensive procedure. Unusually improving foundations by underpinning is a concern for building insurers. A premium is likely to be attached to a house that has been previously underpinned; even though neighbouring properties, which had not been underpinned, would probably be more at risk.

Traditional underpinning involves excavating pits underneath the existing foundations. Only quite small narrow pits can be excavated at any one time, otherwise the building would collapse. The pits are then filled with concrete. Metal bars are set into the sides of the concrete, so that when the next bay is excavated and poured, it will connect to the previous bay via the bars. Once the concrete in a bay has set, it is dry packed tight underneath the original foundations. The next bay can then be excavated, and the process repeated (Figure 4.10.1). A building may be partially or completely underpinned depending on the circumstances.

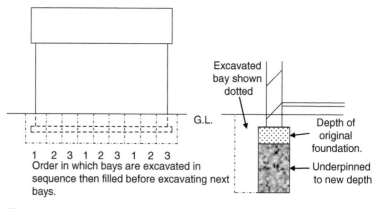

Figure 4.10.1   Underpinning.

Where poor ground conditions dictate, or where the risk of ground shrinkage exceeds about 2.5 metres depth, traditional underpinning becomes impractical. Beyond this depth, piled underpinning is used.

There are several ways to underpin existing buildings.

Mini piling involves drilling or shell auguring small diameter holes underneath the existing foundations. The piled holes are then filled with reinforced concrete. There are a number of ways that this can be carried out.

The piles can be formed both inside and outside the building. The building is then supported on 'needle beams' spanning across piles (Figure 4.10.3). The beams span between the piles, through holes cut into the wall, below ground.

Alternatively, double piles can be formed outside the buildings. Cantilever beams are formed across the piles, set into pockets cut into the walls, below ground (Figure 4.10.2).

A third method involves replacing the floor inside the building with a reinforced concrete raft supported on piles (Figure 4.10.4). Pockets are formed in the existing walls below ground. The reinforced concrete raft is built into the pockets to provide support to the walls.

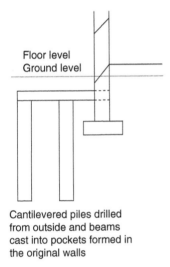

Floor level
Ground level

Cantilevered piles drilled
from outside and beams
cast into pockets formed in
the original walls

Figure 4.10.2 Cantilever beam and piles.

In addition to the methods described, there are also other methods. The piles can be drilled in diagonally. Alternatively, hydraulic jacks can be used to drive sectional precast piles into the ground from directly under the existing foundations.

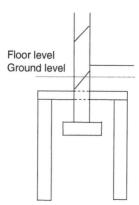

Floor level
Ground level

Piled either side and needle
beams inserted through
holes formed in the original
walls

Figure 4.10.3   Needled through and piles.

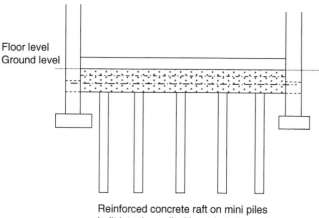

Floor level
Ground level

Reinforced concrete raft on mini piles
built into the wall with pockets formed
in the existing walls

Figure 4.10.4   Piled raft and pockets.

4.11

Expanding Foam Underpinning

Holes are drilled into the ground under floors or foundations. Expanding resin foam is injected into the ground. The expansion can lift floors back into place. The foam then hardens and provides support.

Holes are drilled into the ground. Expanding resin foam is injected underneath the foundations or ground bearing floor slab. The pressure generated by the expanding foam is used to lift the floor or walls back up into place, in some circumstances (Figure 4.11.1). The foam then hardens and provides support. This type of repair is particularly suited to ground bearing floor slabs. The pressure generated under the floor slab can lift it back into place.

Practical Guide to Diagnosing Structural Movement in Buildings, Second Edition.
Malcolm Holland.
© 2023 John Wiley & Sons Ltd. Published 2023 by John Wiley & Sons Ltd.

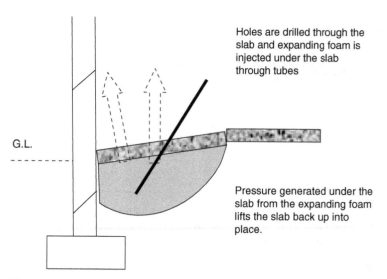

Figure 4.11.1   Injected expanding foam underpinning.

4.12

Grouting

Grouting is used to fill voids. A cement or lime-based slurry is introduced into walls through tubes. The slurry cures to form a solid mass. It can be used to stabilize old stone walls, old drains or mine workings.

Grouting is used to fill voids. It is used to stabilize ground bearing floor slabs or rubble filled walls (Figure 4.12.1). Traditionally the grout was gravity fed into the walls. It may now be introduced under pressure. A cementitious or lime-based slurry is introduced through holes drilled into the walls or floor slab. The slurry fills any voids. The slurry then cures to form a solid mass.

Grouting can also be used to fill and stabilize old mine workings or underground voids, caused by brine pumping or other activities. It can also be used to fill redundant drainpipes or sewers, which might otherwise collapse.

Practical Guide to Diagnosing Structural Movement in Buildings, Second Edition. Malcolm Holland.
© 2023 John Wiley & Sons Ltd. Published 2023 by John Wiley & Sons Ltd.

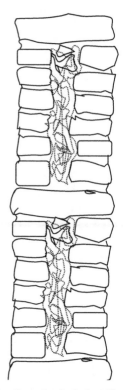

Section through a typical rubble filled
stone wall.

Figure 4.12.1   Grouting a rubble-filled solid stone wall.

4.13

Root Barriers

> *A root barrier is an underground wall which provides a barrier between the roots of a tree and a building. Traditionally this would be a trench filled with concrete but now might be a geo-textile membrane.*

A root barrier is a formed by digging a trench in between a tree and an adjacent building. The trench cuts through any roots. Traditionally, this would then be filled with concrete. Now a proprietary geotextile membrane might be inserted. This prevents the roots from growing towards the threatened building. It also helps to prevent moisture extraction from the soil adjacent to the building.

The root barrier can be set out to intersect a line taken from the trunk of the tree, or trees, to the walls of the building (Figure 4.13.1).

Practical Guide to Diagnosing Structural Movement in Buildings, Second Edition.
Malcolm Holland.
© 2023 John Wiley & Sons Ltd. Published 2023 by John Wiley & Sons Ltd.

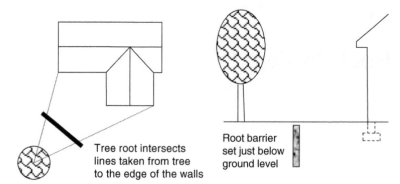

Tree root intersects
lines taken from tree
to the edge of the walls

Root barrier
set just below
ground level

Figure 4.13.1   Root barrier.

Index

Note: Page numbers followed by "f" refers to figures.

Practical Guide to Diagnosing Structural Movement in Buildings, Second Edition.
Malcolm Holland.
© 2023 John Wiley & Sons, Ltd. Published 2023 by John Wiley & Sons, Ltd.